Basic Instrumentation
Lecture Notes and
Study Guide
Volume 2

# PROCESS ANALYZERS
# AND RECORDERS

Basic Instrumentation Lecture Notes
and Study Guide

# VOLUME 2
# PROCESS ANALYZERS AND RECORDERS

**Third Edition**

# RALPH L. MOORE

 INSTRUMENT SOCIETY OF AMERICA

PRENTICE-HALL, INC.
ENGLEWOOD CLIFFS, NJ 07632

**BASIC INSTRUMENTATION LECTURE NOTES AND STUDY GUIDE**
**VOLUME 2: PROCESS ANALYZERS AND RECORDERS**

Copyright © by Instrument Society of America 1982

Printed in the United States of America.

*The Instrument Society of America*
*67 Alexander Drive, P.O. Box 12277*
*Research Triangle Park, NC 27709*

*This Prentice-Hall edition published 1983.*

Library of Congress Catalog Number: 82-81083
ISBN 0-87664-677-1

The Instrument Society of America wishes to acknowledge the cooperation
of those manufacturers, suppliers, and publishers
who granted permission to reproduce material herein
The Society regrets any omission of credit that may have occurred
and will make such corrections in future editions

ISBN (Prentice-Hall Edition): 0-13-062489-6

# FOREWORD

This Study Guide has been an ISA best seller since the first edition was released in 1960 under the tutelage of Education Committee members Joseph E. Casey and Harry Furry. By 1972, feedback from users indicated that (1) Descriptive text material was needed to further amplify the physical, electrical, and chemical principles underlying the operation of each sensor; (2) the contents required expansion to include instruments developed since the original publication; and (3) the contents should be expanded to include instrumentation other than that of the fluid processing industries. With the encouragement of Education Committee Director Warren Carlson and ISA President Walter Bajek (1974), I brought out a greatly expanded edition in 1976 which encompassed the foregoing suggestions. Unfortunately, the volume of text material plus the new sensor concepts required deletion of two subjects that were contained in the first edition—recorders and quality control analyzers.

This Third Edition is again a major expansion to include new information on existing sensors and to include new subjects, e.g., fiber optics, which are described as temperature sensors as well as transmission devices in the chapter on temperature. A new format with the outline and the text in parallel on facing pages greatly expedites use of the Guide. These improvements have necessitated publication of the Third Edition in two volumes, with the fundamental measurements in a companion volume. *Basic Lecture Notes and Study Guide — Measurement Fundaments.*

A new chapter on recorders has been included in this volume. We appreciate the cooperation of manufacturers who generously supplied detailed information and illustrations of their recorders, and we gratefully acknowledge the contributions of Thomas MacGill, who did the preliminary writing on several sections of the chapter.

While a wealth of new material has been added, the philosophy of former editions has been retained. Emphasis continues to be on qualitative descriptions of the operation of sensors; numerous pictures and diagrams of the instruments being described; a minimum of mathematical equations included only where they are required for illustrative purposes; and the inclusion of application information that illustrates the advantages and limitations of the device being described.

Finally, I wish to thank the many ISA staff members at the Research Triangle Park, N.C., headquarters for their suggestions, guidance, and cooperation; and to thank Mrs. Beverle Eastburn for her invaluable help with the manuscript preparation.

Ralph L. Moore
Wilmington, Delaware
January 21, 1982

# ACKNOWLEDGMENTS

The author wishes to thank the following companies who have given permission for their drawings to be used in this text. If any company names have been omitted, the oversight will be corrected in subsequent editions.

Baily Meter Company
Beckman Instruments
Bristol Babcock
Esterline Angus
Fischer & Porter
Fisher Controls
John Fluke Manufacturing Company
Foxboro Company
General Scanning
Gould, Inc.
Honeywell, Inc.
Moore Products Company
Sybron/Taylor
Texas Instruments

# CONTENTS

# CONTENTS (continued)

# 1

# INTRODUCTION TO MEASUREMENT FUNDAMENTALS
## CONCEPTS OF MEASUREMENT
## REFERENCES

## CONCEPTS OF MEASUREMENT

Measurement has been defined as the extraction of signals, which represent parameters or variables (1)*, from physical and chemical systems or processes. The performance of a control system can never surpass that of its associated measuring devices. A good example is a human being. Information about his surroundings is acquired through his senses (sight, smell, hearing, touch, taste). The information is converted into electrical impulses and passed on to the relevant part of the brain, which processes the information and decides the next move to make toward the objective (2). This is called a closed-loop operation. Performance will deteriorate as the senses become impaired, e.g., the ability to drive an automobile diminishes as your ability to see decreases. On the other hand, human performance can be improved if the senses are extended. Thus, the ability to drive is increased by corrective glasses. The purpose of a measuring device is to extend the human senses.

---

*Numbers in parenthese refer to similarly numbered sources in *References* at end of the section.

The output of a measuring device is information concerning the state of the process to which it is connected. The information output is classified relative to some previously defined point of reference. A measuring instrument has its output compared to a precise reference known as a standard. A standard is an arbitrarily chosen reference of suitable magnitude that is assumed to be unvarying (2).

Measurement engineering is a recognized discipline (3). The first principle of this science is: A valid measurement is made when the amount of information obtained is maximized while the amount of energy being taken from the process to obtain the measurement is minimized. A good example of this principle can be found by studying the strain gage.

A stiff strain gage alters the stress field of a test piece. Generally, as energy is drawn from a process, the process is altered, and the measurement is not strictly correct. A basic technique in making measurements is to minimize the alteration of the process, and thus reduce the effect on the measurement to a level of insignificance. This technique can be used in most appliations, and the measurements are valid for all practical purposes.

Energy transfer is involved in all measuring systems. Transducer is a general term for an energy conversion device. A mercury thermometer is a transducer that converts temperature into an equivalent length. Sensing transducers are classified as follows:

(1) Self-generating transducers are those that produce an output from a single energy input. A thermocouple is an example of this type of tansducer.

(2) External power transducers are those that require two energy inputs to provide an output. An example is a resistance thermometer, which requires an energy input from the quantity to be observed and a second electrical input to the measuring bridge (3).

The purpose of a measurement is to obtain the true value of a process quantity by comparing it to a standard or reference. A measurement system can include: comparator, reference, an amplifying element, and a transducer. Errors are reduced with proper application of the instrument and by compensation. The actual value of the measurement is the value of the process quantity, obtained by an instrument, with no systematic or random errors (4). The following definitions apply to measuring devices (11).

**accuracy.** Conformity of an indicated value to an accepted standard value, or true value (5).

**repeatability.** The closeness of agreement among a number of consecutive measurements of the output for the same value of input under identical operating conditions

**resolution.** The smallest interval between two adjacent discrete details that can be distinquished one from the other (6).

**hysteresis.** The maximum difference obtained as an output for the same input between the upscale and downscale values during a full range transverse in each direction (5).

**sensitivity.** The ratio of change in output to the change in input magnitudes.

**precision.** The higher the precision the narrower the intermediate range between two reliably different states (2).

**reproducibility.** The ability of a system of elements to maintain its output/input precision over a relatively long period of time (6).

The definition of accuracy presents a quandry. Accuracy is a relationship to a true quantity. To demonstrate its presence appears to surpass human ability (8). Absolute accuracy is thus unobtainable. However, a definition of accuracy can be approached under highly controlled conditions such as those that exist at the National Bureau of Standards (NBS). Thus the "true" value is a calibration standard with a known relationship (traceable) to NBS. in a large number of installations, however, it is repeatability (not accuracy) that is required, significantly reducing the care with which calibration is performed.

The application of measuring devices requires consideration of four questions (9, 10):

1. What are the characteristics of the process that the instrument will measure?

2. How well does the measurement represent the characteristic or condition being measured?

3. What does the measurement indicate with regard to actual process operation?

4. How is the instrument output going to be used?

How well the measurement represents the true value (discussed above) of the measured quantity is affected by a number of factors. Take temperature, for example. A question exists: Is there one measureable temperature in a process that is representative (9)? Stratification, dead pockets, hot spots, and other conditions can result in significant differences in temperature for different measurement points. It is desirable to eliminate these differences by carefully choosing the most representative measuring point, and then recognize that it is less than truly representative.

Instrument accuracy is specified on a steady-state basis. Dynamic performance (that is, how the measurement varies with respect to time) is of equal importance. A temperature measurement system is again a good example. Thermal elements are almost always installed in thermowells to protect them from the process fluids and to make withdrawal possible without interrupting process operation (9). The thermowell introduces a considerable time lag, whereupon a bare thermal element in a high-velocity fluid stream may respond to a temperature change in a few seconds, while the same element in a thermowell and a low-velocity stream can require minutes to respond.

The outputs of measuring elements on a process can be considered to be time-varying values of product properties (10). There are three general classes of process output. The one that yields the most complete description of data is the analog representation, which is a point-by-point variation in time of each product property. However, information in such detail may be unnecessary, or even undesirable. In this case, the "shorthand" statistical parameters of mean and standard deviation are used to describe a large quantity of process data. The third representation is based on the frequencies occurring in the input signal and is called the power density spectrum (10).

Formerly, instruments were considered as an "add-on" item to the producing plant. Measurements are now becoming a major factor in reducing rather than increasing the overall plant investment (9). An example is automatic custody transfer in the petroleum industry. Manual gaging has been replaced with automatic blending systems that remarkably reduce tankage requirements as well as increasing accuracy. Modern measuring instruments can be designed and installed to provide better reliability, as well as increased accuracy, then the gaging stick and the manual sampling techniques.

Three factors enter into the selection and installation of measuring devices: significance of the measurement, proper application, and cost. Sound technical judgment based on these factors is required for the selection of necessary and appropriate measurements for efficient plant operation (9).

# REFERENCES:

1. Krigman, A., "ICON," *Instruments and Control Systems,* April, 1971.
2. Beaven, C.H.J. and N.G. Maroudas, "Systems Engineering in the Process Industries - No. 2, Measurement," *British Chemical Engineering,* April, 1966.
3. Stein, P.K., "Measurement Engineering," *Instruments and Control Systemns,* April, 1964.
4. Born, G.J. and E.J. Durbin, "Theory of Measurement," *Instruments and Control Systems,* November, 1968.
5. *Process Measurement and Control Terminology,* SAMA Standard PMC 20-2-1970.
6. *Terminology for Automatic Control,* American Standard ASA C85.1-1963.
7. Kemp, R.D., "Accuracy for Engineers," *Instrumentation Technology,* May, 1967.
8. Merewether, E.K., "Accuracy — Instrumentation's Holy Grail," *ISA Journal,* January, 1965.
9. Howe, W.H., "Effective Selection of Measurements for Process Control," *Process Control Handbook,* Instruments Publishing Co.
10. Chope, H.R., "Instrument Dynamics for On-Line Measurements," *ISA Journal,* September, 1963.
11. Deutsch, W.G., "Precision, Accuracy, and Resolution," *ISA Journal,* August, 1965, p. 85.

# 2
# DENSITY MEASUREMENT

## INTRODUCTION

Density has been described as the proportionality factor between volumetric inventory and mass (weight) inventory. As might be expected, density is measured by many of the same instruments used in the measurement of inventory.

If volumetric inventory is constant, then density can be measured by the diaphragm-type transmitter, the bubbler system, and the differential pressure transmitter, all of which are described in Volume 1, Chapter 3, "Level Measurement," with the static weighing devices described in Volume 1, Chapter 4, "Weight Measurement." Nuclear radiation gages are also used to measure density under these conditions.

I. Density
  A. Principles of measurement
    1. Pressure at bottom of fluid proportional to density
      a. Liquid level meter
        (1) Measuring tank
        (2) Reference tank
        (3) Manometer — pressure differential measurement
    2. Weight of given fluid proportional to density
      a. Displacement meter (see Fig. 2-1)
        (1) Chamber
        (2) Displacer and balance beam
        (3) Pneumatic servomechanism and recorder
      b. Hydrometer (see Fig. 2-2)
        (1) Chamber — inlet, outlet, overflow tubes
        (2) Hydrometer
        (3) Armature and inductance coils
        (4) Inductance bridge recorder
  B. Factors affecting measurement
    1. Temperature
      a. Manual correction
      b. Automatic compensation — thermostatic heating unit
    2. Barometric pressure

II. Chain-balanced float density transducers (see Fig. 2-3)
  A. Theory of measurement
    1. Measures density by buoyancy of totally submerged float and chain
    2. Float weighted to support half the weight of the chain
    3. Density increase causes float to rise, supporting more chain
    4. Density decrease causes float to fall, supporting less chain
    5. Float position transmitted to a linear variable differential transformer
  B. Operating conditions
    1. Maximum pressure — 200 psi
    2. Maximum temperature — 100°C; temperature compensation provided
    3. Maximum viscosity — 50 cp
    4. Range — 0.6 to 3.5 SG
  C. Applications
    1. Wide range of liquids
      a. Pumped sample
      b. Pipeline
      c. Gravity flow

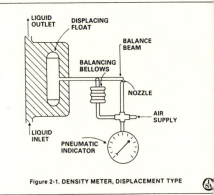

Figure 2-1. DENSITY METER, DISPLACEMENT TYPE

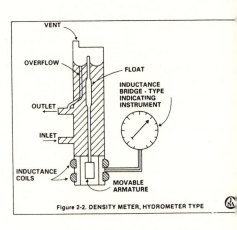

Figure 2-2. DENSITY METER, HYDROMETER TYPE

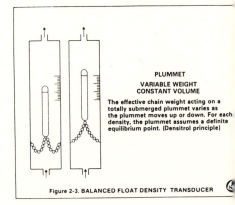

PLUMMET
VARIABLE WEIGHT
CONSTANT VOLUME

The effective chain weight acting on a totally submerged plummet varies as the plummet moves up or down. For each density, the plummet assumes a definite equilibrium point. (Densitrol principle)

Figure 2-3. BALANCED FLOAT DENSITY TRANSDUCER

The *tubular float displacer* can be used to measure density much as it is used to measure level. Archimedes' Law is again applied to the fully submerged displacer. The submerged displacer is accurately weighed. The displacer is buoyed up, or shows a loss of weight, by a force equal to the liquid displaced. But since the volume of the liquid displaced will be a constant equal to the volume of the displacer, the buoying force will be proportional to the density of the surrounding liquid. In practice, the displacer is normalized by filling it with liquid. Thus it assumes a neutral position when the density of the surrounding liquid is the same as that of the fill. In this way, specific gravity is actually measured. Further, temperature compensation can be proved by filling the displacer with the same fluid as that of which the density is being measured (7).

The *hydrometer* (1) is widely used for the determination of liquid density. It is a weighted float that is manually dropped into the liquid. The buoyancy of the liquid causes the float to sink to a depth proportional to the liquid density. The measurement is indicated on a scale inscribed on the upper portion of the float.

The hydrometer has been developed into a continuously measuring and indicating instrument (2). One model of this instrument has the weighted float enclosed in a transparent Pyrex tube, which can be connected directly into the pipeline. The weighted float is completely submerged in this case. However, a density increase causes the float to rise and a decrease causes it to sink; thus a calibrated scale on the Pyrex enclosure can be calibrated in terms of density. Another design (3) uses an opaque float, which interrupts the beam from a light source to a photocell. The amount of light reaching the photocell is proportional to density.

A *chain-balanced float density transducer* also uses the weighted float inside a flow chamber. The float is connected to the chamber by chains. The effective chain weight acting on the totally submerged float varies as the float moves up and down, and for each density the float assumes a new position (4). Float position is then measured by a linear variable differential transformer, which provides a millivoltage output proportional to float position and, thus, to density.

III. Vibrating U-tube transducer (see Fig. 2-4)
    A. Theory of measurement
        1. Density is proportional to mass contained in the U-tube
        2. Vibration of U-tube is proportional to mass
        3. U-tube vibration is sensed by voice coil
    B. Operating conditions
        1. Maximum pressure — 1000 psi
        2. Maximum temperature — 300°F; temperature compensation provided
        3. No maximum viscosity limit
        4. Range — 0.5 to 2.0 SG
    C. Applications
        1. Liquids, pulps, slurries
            a. Pumped sample
            b. Pipeline installation
            c. Gravity flow

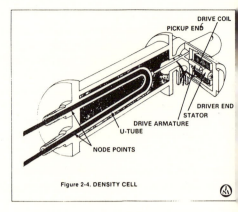

Figure 2-4. DENSITY CELL

IV. Direct weighing U-tube transducer (see Fig. 2-5)
    A. Theory of measurement
        1. Density is proportional to weight contained in the U-tube
        2. Weight of fluid in U-tube changes the nozzle-flapper relationship
        3. Change in flapper position causes change in pneumatic backpressure, which is proportional to density change
    B. Operating conditions
        1. Maximum pressure — 50 psi
        2. Maximum temperature — 200°F
        3. Range 1 to 2 SG
        4. Accuracy ±1%

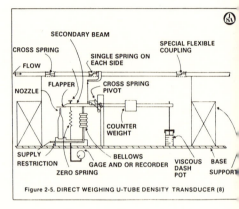

Figure 2-5. DIRECT WEIGHING U-TUBE DENSITY TRANSDUCER (8)

V. Vibration-type density sensor (see Figs. 2-6 and 2-7)
    A. Theory of measurement
        1. If a container of fluid (Fig. 2-6) suspended on a spring is displaced, the frequency of the resulting vibration when released is varied with the density of fluid
        2. An elastic member, stretched tight and constrained, will also vibrate at a frequency that varies as the fluid surrounding the elastic member
        3. A detector is placed near the elastic member, and senses its displacement. The detector transmits a signal at the frequency of the elastic member and proportional to displacement

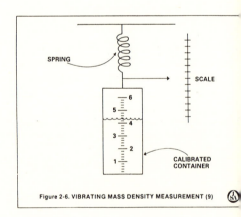

Figure 2-6. VIBRATING MASS DENSITY MEASUREMENT (9)

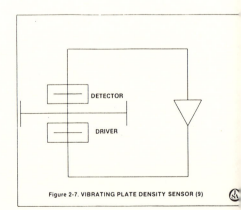

Figure 2-7. VIBRATING PLATE DENSITY SENSOR (9)

A *U-tube piping* configuration is also used in density measuring instruments (5). The U-tube and its contents are driven into mechanical vibrations by an electrically excited drive coil. The vibration is a function of the mass of the material contained in the U-tube. It is sensed by a pickup coil, consisting of an armature and coil. Thus the vibration introduces an ac voltage in the coil. The output of the coil is then converted into a millivoltage output of the transducer.

The U-tube configuration is also used with a pneumatic *flapper-nozzle* type of transducer. The U-tube is mounted in a manner that permits it to flex with the weight of material contained in it. Since the volume contained is constant, the weight is proportional to density. A flapper is attached to the U-tube, and as the tube moves with varying weight, the flapper moves in relation to a pneumatic nozzle. The pneumatic backpressure on the nozzle is thus proportional to density (8).

Another type of vibration densitometer is the *distributed spring mass type* (9). It consists of an elastic member constrained to vibrate at a fundamental resonant frequency. An example is the common drum used in a dance band or a marching band. When hit, the drum head vibrates — and the quality of sound heard depends on the density of the air surrounding the drum. Intuitively, the resonant frequency decreases when fluid density increases. In practice, a means of maintaining the plate at its resonant frequency is provided. A detector is positioned such that it senses the mechanical displacement of the plate and produces an electrical signal that is proportional to the displacement. This signal is then amplified and furnished to an oscillator driver, which in turn imparts a mechanical driving force to the plate at the same frequency as the driving oscillation. The plate experiences the largest displacement when driven at its resonant frequency. Thus the feedback circuit from detector through the amplifier to the oscillator will force the plate to oscillate at its maximum displacement, or its resonant frequency. And the resonant frequency will vary with the density of the surrounding fluid.

    4. This circuit will cause the member to oscillate at its resonant frequency, which varies with the density of the surrounding fluid

  B. Advantages
    1. Used with liquids, gases, and slurries
    2. Small size
    3. Fast response
    4. Easily cleaned for food processing

VI. Float meter for gas density (see Fig. 2-8)
  A. Theory of measurement
    1. Gas sample introduced into float chamber
    2. Float chamber weight measured by calibrating weights
    3. Varying gas density changes weight of gas in the float, causing it to move vertically
    4. Vertical movement measured and calibrated in terms of density

  B. Service conditions
    1. Maximum pressure — 200 psi
    2. Maximum flow — 16 scfh

  C. Advantages
    1. Measures specific gravity directly
    2. Motion balance mechanism records directly

  D. Disadvantages
    1. Complex mechanism
    2. Must be compensated for ambient conditions

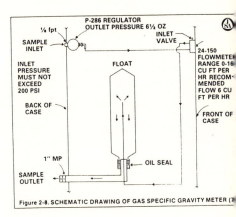

Figure 2-8. SCHEMATIC DRAWING OF GAS SPECIFIC GRAVITY METER (7)

VII. Gas density balance (see Fig. 2-9)
  A. Theory of measurement
    1. A dumbbell pivoted in the sample chamber has one ball punctured; the unpunctured ball rises and falls with gas density in the chamber
    2. Dumbbell position is measured
    3. Two electrodes establish an electrostatic field around the dumbbell
    4. Dumbbell is forced back to its original position by increasing electrostatic force on appropriate electrode
    5. Electrode voltage varies as gas density

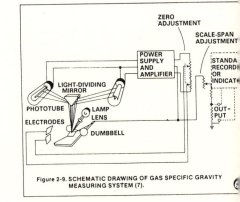

Figure 2-9. SCHEMATIC DRAWING OF GAS SPECIFIC GRAVITY MEASURING SYSTEM (7).

VIII. Other applications
  A. Viscous drag gas density meter
  B. Dynamic densitometer
  C. Piezoelectric crystal densitometer
  D. Dynamic densitometer for gases
  E. Boiling point elevation
  F. Radiation gage

Several density-measuring devices are designed specifically for use with gases. One is a *float-type element* that operates on the comparison of the weight of a volume of gas to that of an equal volume of air (7). A sample of gas is introduced into a hollow, cylindrical float. Calibrating weights are adjusted until the pen arm linkage system attached to the float reads the density of the gas. The vertical position of the float will then depend directly upon the density of the sample of the gas that supports a portion of the weight of the float. A change in the density of the gas in the float causes the float to move vertically, and the vertical movement is proportional to gas density.

Another densitometer designed for gas density measurement is the continuous gas density balance (8). A light dumbell is supported by a horizontal quartz fiber inside a measuring cell. One ball of the dumbbell is punctured, making it insensitive to buoyancy effects. The other ball rises or dips with the density of the surrounding gas. It is coated with rhodium to make it electrically conductive. Two electrodes external to the cell establish an electrostatic field around the ball. A varying potential on the electrodes subjects the ball to an electrostatic force. Thus, as changing gas density causes the dumbbell to move, the force causing movement can be counterbalanced by the electrostatic force. The potential on the electrodes can be chosen to make the ball remain stationary or to null it. The potential is therefore a measure of density.

A *viscous drag gas density meter* has two separate chambers, one filled with air and the other with a gas sample. Driven impellers in each chamber impart opposite rotation in the gas columns. Nonrotating impellers in the chambers are coupled together to measure relative drag, which can be calibrated in terms of the density of the sample gas (9).

A *dynamic densitometer* measures the hydraulic coupling between a rotating impeller and restrained stator or turbine in the flow pipe line. If the impeller rotation is at a constant angular velocity, the coupling between impeller and turbine will be directly proportional to the density of the flowing fluid (6).

The *piezoelectric crystal densitometer* utilizes the sensitivity of a piezoelectric crystal at its resonant frequency to the density of a fluid surrounding the crystal. Thus output voltage is proportional to density. Unfortunately, output voltage is also proportional to the velocity of sound in the liquid. Compensation is required for the change in the velocity of sound with temperature and pressure.

A *dynamic densitometer for gases* operates on the fan law, which says that pressure rise across a fan or blower is directly proportional to the density of the flowing fluid as the rotational speed of the fan or blower is maintained constant. A differential pressure meter can then be used to measure density (6).

In some cases, *boiling point elevation* can be calibrated in terms of liquid density. The temperature of the boiling sample is compared to that of water at the same pressure. For a particular liquid, boiling point elevation can be calibrated in terms of density at standard temperature.

The *radiation gage* described in the chapter on "Level Measurement" can also be used to measure density. It is widely employed to measure the solids per unit volume in slurries.

# REFERENCES

1. Princo Instruments Co., Bulletin 6-11-69.
2. Ibid., Bulletin No. W-6-70.
3. Ess Instrument Co. Bulletin.
4. Princo Instruments Co., Bulletin W-7-70.
5. Automation Products Inc., Bulletin J67D.
6. Halsell, C.M., "Mass Flowmeters, A New Tool for Process Instrumentation," *ISA Journal,* June 1960, p. 55.
7. Carroll, G.C., *Industrial Process Measuring Instruments,* McGraw Hill, New York, 1962, Chapter 7.
8. Halliburton Co., Bulletin SP 11002, 1971.
9. ITT Barton, "The Theory and Operation of Vibration Type Densitometers," Technical Information Bulletin No. 13-G1-56, 1971.

# 3
# HEAT FLUX MEASUREMENT SENSOR TYPES

## INTRODUCTION

A difference in pressure causes fluid to flow, and that flow is measured with a flowmeter, as described in the chapter "Flow Measurement" in Volume 1. A difference in voltage will cause electron flow, called current, which is measured with an ammeter. Similarly, a difference in temperature will cause heat to flow, and the flow is called heat flux. Surprisingly, only a few transducers are available to measure heat flux, and they have been developed since the 1950s. Reflection on the complexities of heat transfer and the effect of those complexities on the measurement explain the relatively late development of these sensors.

The chapter on temperature measurement in Volume 1 defined temperature as a measure of the heat in a body. Further, heat will flow from a point of high heat intensity to a point of lower intensity. Since heat intensity can be measured by temperature, the heat flow is described by Equations 6-1 and 6-2 in that chapter:

$$\frac{\Delta T_s}{\Delta t} = \left[\frac{1}{Mc}\right] \left[Q_f - Q_c - Q_f\right] \qquad (6\text{-}1)$$

where:

$\dfrac{\Delta T_s}{\Delta t}$ = the change in temperature with time

$M$ = the weight of the fluid in which temperature is being measured

$c$ = specific heat of the fluid

$Q_f$ = heat transferred from fluid to sensor

$Q_c$ = heat transferred from the sensor by conduction

$Q_r$ = heat transferred from the sensor by radiation

and

$$Q_f = hA\,(T_f - T_s) \qquad (6\text{-}2)$$

where

$h$ = heat transfer coefficient

$A$ = area of heat transfer

$T_f$ = temperature of fluid

$T_s$ = temperature of sensor

As shown in Equation 6-1, heat is transferred by basically two modes, namely, radiation and convection. Heat transfer by radiation is a completely different physical phenomenon than is heat transfer by convection (1). This difference represents the basic difficulty in heat flux measurement. Radiant heat transfer has been described previously under the subject "Radiation Pyrometers." Convective heat transfer is also described in the chapter in Volume 1 entitled "Temperature Measurement." It is designated by the heat transfer $Q_f$ in Equations 6-1 and 6-2. Equation 6-2 shows $Q_f$ to be dependent on the surface conductance or heat transfer coefficient $h$. This coefficient is in turn dependent on the velocity and state of the fluid flowing past the measuring element, as well as the size and shape of the element itself. When these complexities of heat flux measurement are taken into account, the care necessary for a satisfactory application becomes apparent.

Heat flux measurement is also susceptible to the "instrument effect" discussed in Chapter 1. The sensor, in its operation, may either add heat or absorb heat in the measuring process. Thus the measurement tends to change the quantity being measured. This effect is more troublesome to heat flux measurement than in most other measurements. Thus, in selecting a heat flux transducer, it is necessary to decide what is to be measured and to know what the measurement means.

Figure 3-1 shows heat arriving and leaving the surface in the form of arrows. Some of the energy arriving by radiation ($q_{inc}$) and by convection ($q_{conv}$) will be lost through reflection ($q_{ref}$) and re-radiation ($q_{rr}$). Thus it is apparent that there are several components of heat transfer that must be considered when estimating heat flux through the surface ($q_{net}$) from the sensor output. Ultimately, however, all heat sensors are sensitive only to the net rate of heat transfer to the sensing element ($q_{net}$).

The two major categories of heat flux transducers are calorimeters and radiometers. Calorimeters are used for measuring total heat transfer (radiant plus convective), while radiometers are used in measuring radiant heat transfer only. An optical window covering the sensor makes the radiometer insensitive to convection. The sensors used in both calorimeters and radiometers are commonly of five basic types, outlined on the following pages.

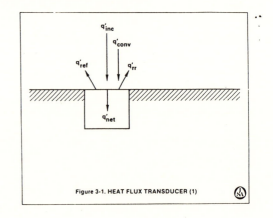

Figure 3-1. HEAT FLUX TRANSDUCER (1)

I. Slug-type heat flux sensors
   A. Heat transfer measurement (see Fig. 3-2)
      1. Theory of measurement
         a. Transferred heat causes slug temperature to rise
         b. Rate of temperature rise is proportional to heat transfer
      2. Advantages
         a. Simplicity
         b. Fast response
         c. Theoretically, no calibration required
      3. Disadvantages
         a. Data difficult to interpret
         b. Not used for time-varying measurements
   B. Convection heat transfer coefficient measurement (see Fig. 3-3)
      1. Theory of measurement
         a. Sensing plug and guard cup are heated by heating coils
         b. Temperature of plug and cup are monitored by thermocouples
         c. Electrical power is applied to both cup and plug to maintain zero temperature difference
         d. Energy transferred to plug from coil is transferred to the environment
      2. Coefficient range is 5 to 40 Btu/hr-ft²
      3. Material of construction
         a. Sensing plug — highly conducting
         b. Guard cup — highly conducting
         c. Support case — insulating
         d. Heating coil — Bifilar wound
         e. Thermocouple — iron constantan

II. Thin disc sensors
   A. Theory of measurement (see Fig. 3-4)
      1. When heat flux into disc surface is uniform, temperature difference between center and edge of disc is proportional to heat flux
      2. Differential thermocouple between center and edge of disc measures flux
   B. Characteristics
      1. Temperature difference of 360°F results in 10 mV output
      2. Range of from 5 to 3000 Btu/ft²-sec
      3. Time constant of from .05 to .50 second
      4. Maximum temperature of 450°F
   C. Material of construction
      1. Disc of constantan

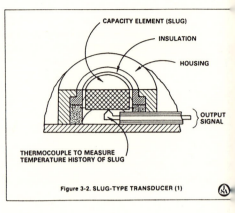

Figure 3-2. SLUG-TYPE TRANSDUCER (1)

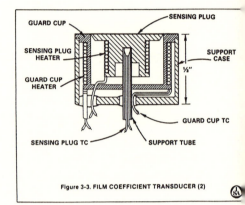

Figure 3-3. FILM COEFFICIENT TRANSDUCER (2)

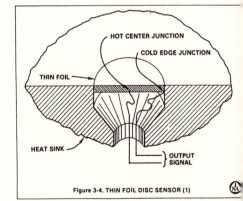

Figure 3-4. THIN FOIL DISC SENSOR (1)

The *slug-type sensor* consists of a metallic disc, or slug, of known weight and thermal mass insulated from its housing and having a thermocouple attached to its back surface. The rate of rise of slug temperature is then proportional to the rate of heat transfer into the slug.

The *thin disc sensor* consists of a thin metallic foil disc bonded at its periphery, over a cavity, to a metallic body. Thermocouples measure the temperature difference between the center and the edge of the foil. The temperature difference is directly proportional to the rate of heat transfer to the foil (1).

The thin disc sensor is widely used in the aerospace industry. It has very rapid transient response. Continuous, time-varying measurements can be easily made.

The millivolt signal produced by the sensor is directly proportional to heat flux. This is desirable in terms of flux measurement, but it does place an upper limit on the heat load that can be accommodated for a given design. Body temperature increases with time as long as the heat flux is positive. The measurement time interval therefore must not exceed the time for the sensor to reach a maximum temperature, as determined by sensor properties. For long-term use or high heating rates, water-cooled instruments are available.

2. Body of copper
3. Copper wires
D. Advantages
   1. Temperature difference proportional to heat flux
   2. Fast response
E. Disadvantages
   1. Head load must be limited or sensor water cooled
   2. Maximum output due to heat flux limitation is about 10 mV

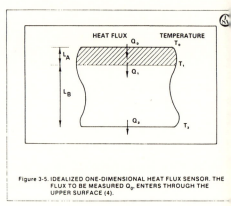

Figure 3-5. IDEALIZED ONE-DIMENSIONAL HEAT FLUX SENSOR. THE FLUX TO BE MEASURED $Q_0$, ENTERS THROUGH THE UPPER SURFACE (4).

III. Thin film sensors
  A. Theory of measurement (see Figs. 3-5 and 3-6)
    1. Very thin temperature-sensitive resistance film changes electrical resistance as it is heated
    2. Electrical resistance change is proportional to surface temperature change due to heat flux arriving at surface
    3. Constant small current applied to sensor will cause measured voltage to change as resistance changes
  B. Construction
    1. Resistance film of platinum
    2. Film deposited on insulating substrate of glass or quartz
  C. Advantages
    1. Extremely fast response
    2. Used in aerodynamic shock tube measurements
  D. Disadvantages
    1. Requires complex analysis to infer heat transfer rate
    2. Not applicable to time-varying heat conditions

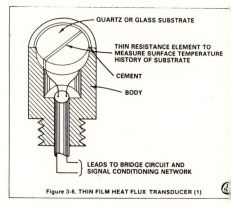

Figure 3-6. THIN FILM HEAT FLUX TRANSDUCER (1)

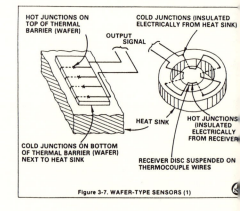

Figure 3-7. WAFER-TYPE SENSORS (1)

IV. Wafer-type sensors
  A. Theory of measurement (see Figs. 3-7 and 3-8)
    1. Heat flux on surface of wafer causes heat flow from surface to heat sink
    2. Heat flow causes a temperature difference proportional to flux
    3. Junctions of thermocouple or thermopile at surface of wafer provide output millivoltage proportional to temperature difference
  B. Characteristics
    1. Wafer thermal conductivity of between .025 and .15 Btu/hr-ft² °F
    2. Wafer thickness of from 0.005 to .05 inch

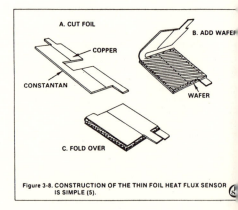

Figure 3-8. CONSTRUCTION OF THE THIN FOIL HEAT FLUX SENSOR IS SIMPLE (5).

The *thin film sensor* is a one-dimensional element in which heat flow is normal to the thermometric surface. The surfaces are instrumented to measure temperature, temperature difference, or time rate of temperature change. The temperature across the film is uniform. The temperature gradient is thus one-dimensional, resulting in the thin film sensor belonging to the broad category of gradient sensors (3).

The thin film sensor is an element that can be considered infinitesimally thin, typically $10^{-6}$-inch thick. Because of the thinness of the film, materials of low density and low specific heat must be used to maximize the temperature difference. The film can be used as a resistance thermometer (4). The low thermal capacity of the film provides response capabilities on the order of microseconds, but still yields high output signals.

Gage output level can be increased by reducing film thickness to raise sensor resistance, but this has the disadvantage of making the gage more fragile. An alternative is to increase excitation voltage, but the higher current increases the self-heating of the sensor. The self-heating raises the film temperature and causes the thermal properties to change.

For one-dimensional heat flow through the film, a constant incident heat flux causes the temperature to rise. Increasing temperature causes an increasing resistance. The sensor thus responds like a resistance thermometer, where a voltage change is measured if a small current is passed through the film (4).

The *thick film sensor* is used when high heating rates, erosion, or ionization are anticipated. It is some 0.01-inch thick. Heat flux is inferred from the time rate of change of mean film temperature, similar to the technique used with the slug-type sensor. Temperature is measured by using the film as a resistance thermometer (3).

The *wafer-type sensor* is thick enough that the junctions of a thermopile are attached to its surfaces and the one-dimensional temperature measured. A thermopile is a number of thermocouples connected in series, making the millivoltage outputs additive. Thus small temperature differences produce large output signals. The hot junctions of the thermopile are attached to the wafer surface, which receives the heat flux, and the cold junctions between the wafer and the substrate (1).

Heat flow causes temperature gradient across the wafer, proportional to the rate of heat transfer. The output of each differential thermocouple is proportional to the temperature drop across the wafer and hence directly proportional to heat flux. Since all sets of junctions are connected in series, the individual outputs are additive, resulting in a strong output signal, which is the average of all the hot and cold junction contributions (1).

3. Wafer density is some 70 lb/ft³
4. Wafer specific heat is about 0.3 Btu/lb-°F
C. Construction (see Fig. 3-8)

V. Suspended disc-type sensor (see Figs. 3-7 and 3-9)
A. Theory of operation
   1. Heat flux incident on one side of the disc causes temperature to rise at surface
   2. Heat flows across disc and from disc to heat sink by the thermocouple or thermopile wires
   3. Thermocouple hot junction is attached to back of the disc, cold junction to heat sink
   4. Temperature difference between disc and sink is proportional to heat flux
   5. Thermocouple provides millivoltage output proportional to heat flux
B. Materials of construction
   1. Disc and heat sink of gold-plated aluminum
   2. Thermocouple wire of iron/constantan

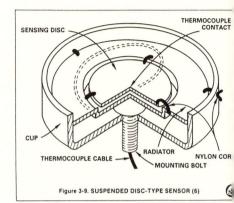

Figure 3-9. SUSPENDED DISC-TYPE SENSOR (6)

The *suspended disc* operates in a fashion similar to the wafer that is bonded to a heat sink, except that it is not bonded, but is suspended by the thermocouple wires. The thermopile hot junctions are attached to the suspended disc, and the thermocouple wires conduct heat from the wafer to the heat sink.

**Heat flux sensor selection.** The complexities of heat flux measurement are apparent from the foregoing description of sensors that make such measurements. As a result, real sensors will have characteristics different from those predicted for ideal devices (3). However, the foregoing descriptions serve as a useful guide in selecting a device for a given application. Selection factors are summarized in Table 3-1.

Table 3-1
Summary of Selection Criteria (3)

1. When the best possible frequency response is required, with moderate or long testing time, the gradient sensor is the best choice.

2. For a given ratio of output temperature to input heat flux, the thin disc sensor has the best frequency response. This is consequently the best choice when heat flux does not strongly depend on sensor temperature, such as in radiation from a high-temperature source.

3. For short measuring times, thin or thick film sensors may be used; the thin film sensor has the greater sensitivity and better frequency response.

4. For the thin film sensor the period of the highest frequency heat flux component of interest may be as small as one-tenth the measurement time. The thick film sensor and wall calorimeter permit accurate measurements only when the period of the highest frequency heat flux component of interest is *not* significantly less than the measuring time.

5. The wall calorimeter should be used only for moderate testing times, since its frequency response is poor. The chief advantage of the wall calorimeter is convenience.

6. For transient measurements using the gradient sensor, accuracy can be improved if inside or outside temperatures are known in addition to the temperature difference.

7. For transient measurement using the thin disc sensor, accuracy can be improved if the center or outside edge temperature is known in addition to the temperature difference.

8. The frequency responses of the gradient sensor and thin disc sensor deteriorate as heat sinks become less perfect. The simpler formulas for heat flow similarly deteriorate.

# REFERENCES

1. Hornbaker, D.R. and D.L. Rall, "Heat Flux Measurements - A Practical Guide," *Instrumentation Technology,* February, 1968, p. 51.
2. Progelof, R.C., et al., "Heat Gage Measures Local Coefficient," *Instrumentation Technology,* December, 1970, p. 75.
3. Baines, D.J., "Selecting Unsteady Heat Flux Sensors," *Instruments and Control Systems,* May, 1972, p. 80.
4. Mason, S.B. and R.L. Varwig, "Increasing the Sensitivity of Thin Film Gages," *Instruments and Control Systems,* May, 1972, p. 105.
5. Hager, N.E., Jr., "Making and Applying a Thin-Foil Heat Flux Sensor," *Instrumentation Technology,* October, 1967, p. 54.
6. Robinson, G.P. and E.D. Crofts, "Sensor for Finding Heat Flux," *Instrumentation Technology,* March, 1967, p. 53.

# 4
# MOISTURE AND HUMIDITY MEASUREMENT

INTRODUCTION
DEWPOINT SENSORS
RELATIVE HUMIDITY SENSORS
ABSOLUTE MOISTURE MEASUREMENT
REFERENCES

## INTRODUCTION

Water content is important in the establishment of environments, whether for human comfort or an industrial process where the presence of too much or too little moisture will vitally affect product specifications (1,2). A multitude of instruments are available for measuring water content (3). As far back as 1881, a historical review by Symonds listed no less than 138 methods and instruments. The tremendous number of devices that have been developed is perhaps a measure of the unsatisfactory nature of the art (3).

Examples of the requirement for humidity control include a lumber-drying kiln, where too much moisture will retard drying and too little moisture will cause case-hardening and honeycombing (1); and paper making, where excess moisture adds to steam costs, may cause losses in strength, and directly reduces profits when the paper is sold by weight (4).

Instruments for measuring water content can be classified as humidity sensors, which are used when water is presented as a vapor in a gaseous medium; and moisture detectors, which are employed when water occurs as a liquid in a solid, liquid, or a gas medium (2).

A number of terms are used in describing water content, some of which are defined in Table 4-2.

Water content-measuring sensors can also be classified as dewpoint types, relative humidity types, and absolute moisture measuring types (6).

Table 4-1
Principles of Moisture and Humidity Sensors (29)

| Principle | How it works | Comments |
|---|---|---|
| Capacitance | Material dielectric constant changes between C plates as $f(H_2O)$. | Gas, liquid, solid; avoid chlorine, ammonia, etc. |
| Condensation | Temperature at which condensate appears on cooled surface is $f(H_2O)$. | Chemically, mechanically, or thermoelectrically; dew or frost. |
| Electrolytic | At known flow rate, electrolysis current is $f(H_2O)$ for gas sample. | If liquid, vaporize or strip to produce gaseous sample. |
| Gravimetric | Weight loss by removing moisture is $f(H_2O)$ in liquid or solid. | $H_2O$ removal be heating, freeze dry, or chemical method. |
| Hygroscopic | Change in length of hair, fiber, etc. absorbing $H_2O$ is $f(H_2O)$. | Simple, easy to calibrate, but lacks precision. |
| Infrared | IR absorption is $f(H_2O)$. Split beam for dry sample comparison. | Favors 1.4 to 1.93 micron; transmit for gas, liquid; reflect for solids. |
| Microwave | Attenuation of radiation is $f(H_2O)$. | Transmit for gas, liquid, solids. |
| Nuclear | Bombard hydrogen atoms, solid sample; $\beta$ particle passage if $f(H_2O)$. | Sensitive to automatic structure, mass density, etc. |
| Piezoelectric | Hygroscopic coating adsorbs water, changes crystal frequency, $f(H_2O)$. | Natural frequency decreases as crystal mass increases. |
| Psychrometric | $\Delta t$ between wet and dry wick thermometers is $f(H_2O)$. | Keep gas (air) at constant velocity for accuracy. |
| RF sensor | RF current due to dielectric change is $f(H_2O)$. | Average sample dielectric constant is 16 times that of water |
| Resistance | Conductivity is $f(H_2O)$ adsorbed. | Avoid gas with dew point $H_2O$. |
| Saturated salt | Temperature at condensation-evaporation equilibrium is dew point. | Use water soluble salt such as lithium chloride (LiCl). |

## Table 4-2
## Definitions (5)

**Total pressure.** Pressure exerted by a mixture of gases or vapors (the sum of the partial pressure of the constituents, by Dalton's Law).

**Partial pressure.** Pressure that a constituent of a gaseous mixture would exert if it alone occupied the volume of the mixture.

**Dew point.** Temperature at which gas is saturated with respect to liquid water.

**Frost point.** Temperature at which gas is saturated with respect to ice.

**Vapor pressure.** Partial pressure of water vapor in a gaseous atmosphere.

**Relative humidity.** Ratio of actual partial pressure of water vapor to saturation vapor pressure at ambient temperature.

**Parts per million by volume.** Ratio of volume of water of dry carrier gas, times $10^5$.

**Parts per million by weight.** Ratio of weight of water vapor to weight of dry carrier gas, times $10^5$.

**Mixing ratio.** Ratio of weight of water vapor to weight of dry carrier gas.

I. Dew point sensors
  A. Condensation type (see Figs. 4-1 and 4-2)
    1. Theory of operation
      a. Gas containing water — vapor is cooled by contact with mirror
      b. As gas is cooled, a temperature will be found at which water vapor condenses on the mirror
      c. The temperature at which condensation occurs is known as the dew point
      d. Mirror temperature is detected by temperature sensor
      e. Condensation is sensed by fog on the mirror
    2. Operating characteristics
      a. Range from –115°F to 125°F
      b. Accuracy ±2°F
    3. Advantages
      a. Wide range
      b. High accuracy
      c. Automated
    4. Disadvantages
      a. High cost
      b. Complexity
      c. High maintenance on mirror
  B. Aluminum oxide impedance type (see Figs. 4-3 and 4-4)
    1. Theory of operation
      a. Coating of gold over porous aluminum oxide substrate
      b. Gold and aluminum form plates of capacitor
      c. Water vapor adsorbed and desorbed by aluminum
      d. Aluminum oxide wall resistance proportional to water adsorbed
      e. Sensor impedance proportional to water vapor pressure
      f. Impedance measured by ac bridge
    2. Characteristics
      a. Dew point range of from –110°C to +60°C
      b. Application pressures of from vacuum to 5000 psig
    3. Advantages
      a. High sensitivity
      b. Wide dynamic range
      c. "In-line" measurement
      d. Wide span
      e. Measure of "frost" point
    4. Disadvantages
      a. Nonuniform calibration
      b. Drift due to aging and contamination

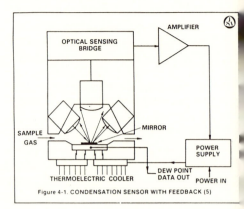

Figure 4-1. CONDENSATION SENSOR WITH FEEDBACK (5)

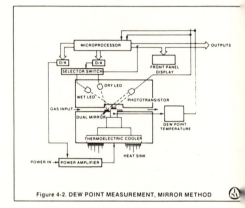

Figure 4-2. DEW POINT MEASUREMENT, MIRROR METHOD

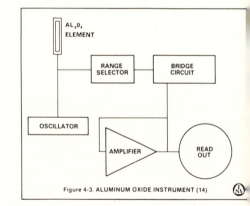

Figure 4-3. ALUMINUM OXIDE INSTRUMENT (14)

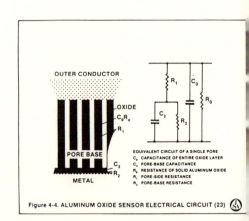

Figure 4-4. ALUMINUM OXIDE SENSOR ELECTRICAL CIRCUIT (23)

Dewpoint-Type Sensors: *Condensation type* (1,2,3,5,7). Some 200 years ago, the determination of dew point by chilled mirror techniques came into use. The temperature of a reflecting surface is slowly reduced until saturation is reached. At that point, the number of water molecules leaving the atmosphere and condensing on the chilled surface equals the molecules leaving the surface and re-entering the atmosphere. The technique represents a fundamental measurement of the dew point temperature, which is the saturation partial pressure of the water vapor. Tables are available for the dew point/vapor pressure relationships.

In measuring the dew point, a temperature sensor is placed in intimate contact with the mirror, and the entire assembly is insulated. Mirror temperature is then used as dew point temperature. Mirror cooling is accomplished using dry ice and acetone, cryogenic fluids, mechanical refrigeration, or thermoelectric techniques. The mirror surface is cooled until it becomes cloudy from the condensing vapor.

Optical detection of the condensate on the chilled mirror is by visual inspection, or by sensors such as photocells and photoresistive devices. While the manually cooled, visually observed optical hygrometer is relatively inexpensive and easy to use, it suffers from not being a continuous measurement and being operator dependent (23, 24).

Modern chilled-mirror hydrometers use thermoelectric cooling devices to chill the mirror surface, while a continuous flow of sample gas is passed over the mirror surface. The mirror is illuminated by a light source and observed by a photodetector bridge network. The mirror, however, is affected by deposition of noncondensable contaminants (particulate matter and chemical salts). The light-scattering properties of the particulate matter cause a gradual increase in the reading of the instrument. A chilled-mirror hygrometer using two mirrors and two photodetectors has been developed for continuous operation. One mirror is cooled to measure dew point; the other is kept at ambient temperature as a reference. Experience has shown, however, that particulate matter accumulates faster on one mirror than on the other. The best results in long-term, unattended operation are achieved by periodically adjusting the balance of light source, mirror surface, photodetector, and electronic control loop on an automatic basis.

Two light-emitting diode (LED) light sources are directed at the chilled and warm (reference) mirrors, respectively. The gas immediately adjacent to the chilled mirror is cooled until the contained water vapor condenses on the mirror surface. As the condensate forms on the mirror, the change in reflectance detected by the optical bridge is developed as a proportional control signal and fed back to drive the thermoelectric cooler. The system periodically switches to the LED directed at the reference mirror, thus providing automatic continuous correction of mirror contaminant buildup.

With the thermoelectric cooler providing just enough cooling to maintain condensate on the chilled mirror, a temperature-measuring element embedded in the mirror measures the dew point temperature (absolute humidity) directly.

*Aluminum oxide impedance type* (14, 23, 25, 28, 29). The thin film aluminum oxide hygrometer sensor is a transducer that converts water vapor pressure into an electrical signal. It consists of an aluminum strip that is anodized to provide a porous oxide layer. A thin coating of gold is evaporated over this structure. The aluminum base and the gold layer form the two electrodes of a capacitor. Water vapor in the gas surrounding the element is adsorbed and desorbed by the aluminum oxide to reach equilibrium. The vapor that equilibrates on the pore walls is functionally related to the vapor pressure of the water vapor in the gas.

The number of water molecules adsorbed on the oxide structure determines the electrical conductivity of the pore wall. Each value of pore wall resistance results in an electrical impedance, which is in turn a direct measure of water vapor pressure. Impedance is measured by an ac bridge measuring device. A constant voltage is applied to the sensor, and the change on impedance of the elements due to a change in water vapor pressure results in a current change. Calibration of current signal with dew point then defines sensor response.

The aluminum oxide sensor provides high sensitivity and a wide dynamic range of moisture measurement but requires frequent calibration because aging and contamination cause long term drifts (28). It can be used with pressures from a few microns of mercury to 500 psig and thus can be used for "in-line" measurements.

C. Saturated heated lithium chloride type (see Figs. 4-5 and 4-6)
  1. Theory of operation
    a. Two parallel wires are wound on a lithium chloride impregnated sleeve that covers a hollow tube
    b. The two wires are not connected, but current flows from one to the other through the lithium chloride coating
    c. The resistance of the lithium chloride varies with its water content
    d. The flowing current heats the lithium chloride, evaporating its water vapor until it reaches equilibrium with its surroundings
    e. At equilibrium, the temperature of the lithium chloride is a measure of the ambient dew point
  2. Characteristics
    a. Range of from –50°F to 160°F
    b. Accuracy ±1.5°F
  3. Advantages
    a. Stability
    b. Sensitivity
  4. Disadvantages
    a. Resistance grid varies with time and contamination
    b. Temperature limitation of 135°F for continuous operation

II. Relative humidity sensors
  A. Psychrometer (see Figs. 4-7 and 4-8)
    1. Principle of measurement
      a. Comparison of wet and dry bulb temperature (wet bulb depression)
      b. Wet bulb measures lower temperatures than dry bulb
      c. Greater cooling effect if air is less saturated
    2. Method of measurement
      a. Psychrometer
        (1) Construction
          (a) Wet bulb — maintained constantly wet (reservoir or wick)
          (b) Dry bulb
          (c) Two-point recording instrument
        (2) Circuitry — resistance thermometer type
    3. Factors affecting measurement
      a. Wet bulb requires air stream of at least 15 feet/second — blower arrangement necessary when below minimum

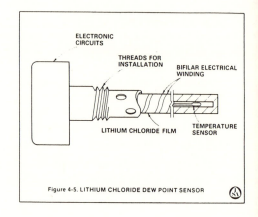

Figure 4-5. LITHIUM CHLORIDE DEW POINT SENSOR

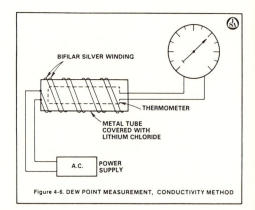

Figure 4-6. DEW POINT MEASUREMENT, CONDUCTIVITY METHOD

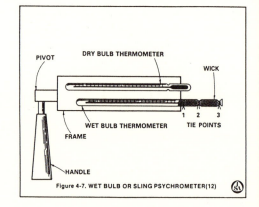

Figure 4-7. WET BULB OR SLING PSYCHROMETER(12)

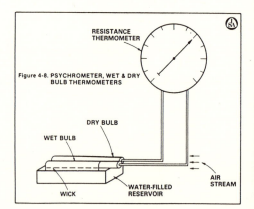

Figure 4-8. PSYCHROMETER, WET & DRY BULB THERMOMETERS

*Heated lithium chloride* (1, 7, 23, 29). The sensor consists of a thin wall tube, generally metal, containing a platinum resistance temperature sensor. The tube is covered with an inert fabric sleeve or wick, which is impregnated with 5 percent dried lithium chloride solution. The entire bobbin is wound with a bifilar winding that is used for heating the assembly. An ac power source is then used to heat the lithium chloride to a point where its ability to absorb is equal to its ability to expel moisture.

The two wires that make up the bifilar winding are not connected, but current flows from one to the other through the intervening lithium chloride coating. The lithium chloride thus acts as a variable resistor between the wires. The resistance is a function of the moisture content of the coating. Current flow is proportional to resistance and is also proportional to moisture content. The current flow causes heating, which in turn causes moisture to evaporate from the lithium chloride. Evaporation continues until moisture equilibrium is attained with the surroundings. The equilibrium temperature is independently measured and related to the dew point of the atmosphere surrounding the element.

Relative Humidity Sensors: *Mechanical type* (1,2,3,5,6,7). Many materials change their dimensions when moisture is adsorbed and desorbed. In 1665, Robert Hooke described this effect on gutstrings. Subsequently, use was made of the expansion of wood, whalebone, paper, whipcord, cotton, silk, quills, animal membranes, and horsehair (3). DeSaussure, in 1783, measured the change in the length of human hair with varying moisture conditions. Since that time, the hair hydrometer has probably been the most widely used instrument for humidity measurement, and it has remained essentially unchanged over the intervening years (9).

Hair and most organic materials adsorb and desorb moisture with their temperatures and the water vapor pressure of the surrounding atmosphere. Human hair increases in length by some 2-1/2 percent as relative humidity increases from 0 to 100 percent. However, exposure to low (under 15 percent) or high (over 85 percent) humidities causes a permanent shift in calibration. Thus the 0 to 100 percent graduations on hair hydrometers are deceiving, since the actual range is from 15 to 85 percent (9). Further, the change in hair length is nonlinear in a logarithmic fashion over the active range (10).

| Range | Sensitivity (Change in length of a 4-inch hair per 1 percent change in RH) |
|-------|------------------------------------------------------------------------------|
| 15% | 0.020 inch |
| 50% | 0.007 inch |
| 80% | 0.00035 inch |

These very small and nonuniform changes in length require mechanical amplification and linearization to be useful. The amplification requires more force than can be provided by a single hair. A number of hairs are mounted in multiples to offset these effects. The mounting then requires uniform spacing for moisture ventilation and uniform tension of all of the hairs.

Relative Humidity Sensors: *Wet/dry bulb psychrometers* (1, 2, 3, 5, 7, 9). The wet/dry bulb concept was observed by Cullen in 1755 (3) wherein he noted that evaporation from a thermometer bulb caused its temperature to fall. The principle was well developed prior to 1900 and forms the basis for the most frequently used sensor (9).

In its simplest form, the psychrometer consists of two mercury-in-glass thermometers, one bare and one with its bulb covered with a piece of moist muslin. The two thermometers are attached to a common base, and air is forced to flow past the bulbs. One way to cause air flow is to swing the thermometer assembly through the air in a circular arc, a technique given the appropriate name of "sling psychrometry." Another technique is to blow air across the bulbs with a small fan. The water on the wick is evaporated into the adjacent air, causing the wet-bulb temperature to fall. It will fall until the evaporation rate required to saturate the air adjacent to the wick is constant. The rate of evaporation depends therefore on the relative humidity of the atmosphere. The wet and dry bulb temperatures can be converted to relative humidity by the use of standard psychrometric tables. The tables, however, are largely experimental, and there is no theory that completely describes the wet-bulb performance (5).

   b. Dry bulb must measure true temperature
   c. Temperature of liquid must be same as dry bulb temperature
   d. Dry-bulb temperatures limited between 32°F and 212°F
  4. Applications of psychrometric measurements
   a. Humidity content
   b. Control of moisture content in gases
   c. Air conditioning
B. Hygrometer (see Figs. 4-9 and 4-10)
  1. Theory of operation
   a. Relative humidity readings based on ability of material to adsorb moisture (hygroscopic)
  2. Types of measurement
   a. Mechanical — linear expansion and contraction of hygroscopic materials
    (1) Hair element — contracts with decrease in humidity
    (2) Mechanical linkage
    (3) Indicating pointer
   b. Electrical — electrical conductivity dependent on moisture content of hygroscopic sensing element
    (1) Sensing element — wire or metal leaf construction, coated with lithium chloride (hygroscopic)
    (2) Electronic recorder — Wheatstone bridge circuit
   c. Chemical color indicators
  3. Applications
   a. Air conditioning systems
   b. Food processing
   c. Textile mills
   d. Lumber industry
   e. Paper industry
   f. Precision instrument manufacturers
C. Piezoelectric (see Fig. 4-11)
  1. Principle of measurement
   a. Hygroscopically coated quartz crystal
   b. Weight changes with water adsorbed
   c. Frequency changes with weight
   d. Moisture is calibrated to frequency
  2. Method of measurement
   a. Two quartz crystals
   b. Hermetically sealed crystal serves as reference
   c. Weight of crystal exposed to moisture changes
   d. Each has separate radio frequency oscillator circuit
   e. Difference in frequencies is measure of relative humidity

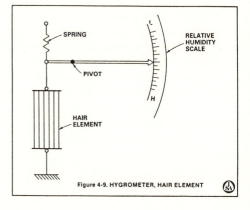

Figure 4-9. HYGROMETER, HAIR ELEMENT

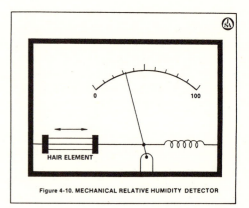

Figure 4-10. MECHANICAL RELATIVE HUMIDITY DETECTOR

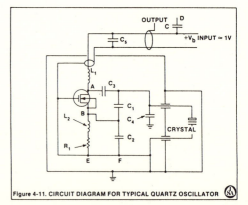

Figure 4-11. CIRCUIT DIAGRAM FOR TYPICAL QUARTZ OSCILLATOR

The accurate performance of a psychrometer depends heavily on both technique and procedures. Technique is discussed by the U.S. Department of Commerce (11). It includes the use of distilled water, clean wicks of the proper lengths and well fitted to the bulb, certified thermometers capable of being read to 0.1°C, and an air flow of at least 15 ft/sec. Procedures include the restriction that the wick be wet only once during any given reading, and that five successive readings of wet and dry bulb be made to assure that the wet bulb reading has reached equilibrium (9).

The care with which readings must be taken is demonstrated by Table 4-3.

Table 4-3
Extent of Error in Psychrometric Readings (12)

| Measured condition | Actual Condition May Be | | |
|---|---|---|---|
| | With 2°F thermometer | With 1° thermometer | With 0.5°F thermometer |
| 67% RH at 50°F | 46 to 93% RH | 56 to 80% RH | 61.5 to 73.5% RH |
| 50% RH at 70°F | 37 to 66% RH | 44 to 56% RH | 47 to 54% RH |
| 90% RH at 70°F | 73 to 100% RH | 81 to 100% RH | 86 to 95% RH |
| 10% RH at 138°F | 7 to 13% RH | 8.5 to 11.5% RH | 9.25 to 10.75% RH |
| 92% RH at 138°F | 82 to 100% RH | 87 to 97% RH | 89 to 94% RH |

The psychrometer is widely used as a local indicating device. Some attempts have been made to develop it into a transmitting sensor by replacing the thermometer bulbs with the hot and cold junctions of a thermocouple (3).

*Piezoelectric sorption* (3, 5, 26, 30, 31). The frequency of oscillation of a hygroscopically coated quartz crystal is decreased when the crystal gains weight due to water sorption on the surface coating. The piezoelectric crystal is used as a sensing element to detect humidity changes. When exposed to an air flow, it produces oscillations related to the moisture content of the air.

A lithium fluoride-coated quartz crystal microbalance makes up the relative humidity sensor (30). A thin layer of coating is added to one surface of a shear-mode oscillating quartz-crystal disc. The crystal surfaces are vacuum deposited on a set of thin metal film electrodes. The electrodes are connected to a feedback amplifier to complete an electronic oscillator. Insensitivity to temperature is attained by choosing the cut angle of the crystal chip with respect to the crystallographic axis so that the frequency-temperature coefficient is zero.

The analyzer operates by monitoring the vibrational frequency change between two quartz crystals connected to two separate radio frequency oscillator circuits. One crystal has the hygroscopic coating and is alternatively exposed to wet and dry filtered process gas (26, 31). The mass and consequently the frequency of the coated crystal changes as water is adsorbed and desorbed on its surface. The other crystal is hermetically sealed and serves as a stable reference frequency. The crystal frequencies are compared via a frequency mixer circuit, and the output is calibrated in moisture content.

Other moisture adsorbing materials are also used for sensing crystal coatings. They include molecular sieves (zeolites) for more sensitive applications or hygroscopic polymers for linearity.

3. Applications
   a. Refineries
   b. Dryer performance
   c. Natural gas production
   d. Liquids separation
   e. Specialty gas production
D. Surface resistivity or conductance
   1. Impedance sensors (see Fig. 4-12)
      a. Theory of operation
         (1) Element of hygroscopic material exhibits large change in impedance with change in moisture content
         (2) Resistance varies with moisture content
         (3) Resistance change is detected by an ac bridge circuit
      b. Construction
         (1) Thin film of hygroscopic material on base material
         (2) Electrodes intermeshed beneath film
      c. Range and accuracy
         (1) Range of 5 to 95%
         (2) Accuracy ±2%
      d. Advantages
         (1) Fast response
         (2) Continuous readout
         (3) Simplicity
      e. Disadvantages
         (1) Temperature sensitive (0.3% RH/°C)

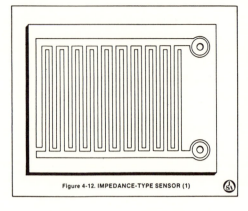

Figure 4-12. IMPEDANCE-TYPE SENSOR (1)

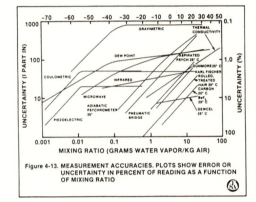

Figure 4-13. MEASUREMENT ACCURACIES. PLOTS SHOW ERROR OR UNCERTAINTY IN PERCENT OF READING AS A FUNCTION OF MIXING RATIO

III. Absolute moisture measurement
   A. Gravimetric sensors (see Fig. 4-14)
      1. Theory of operation
         a. Sample of liquid or solid material is weighed before and after drying in an oven; or dessicant is weighed before and after gas is passed through
         b. The difference in the two weights is the amount of moisture in the sample
      2. Range and accuracy
         a. Range 0.1 to 100% moisture
         b. Accuracy ±0.2% (in laboratory)
      3. Advantages
         a. High accuracy
         b. Applies to liquid, solid, or gas
      4. Disadvantages
         a. Precise technique required

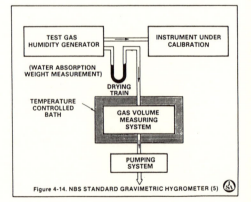

Figure 4-14. NBS STANDARD GRAVIMETRIC HYGROMETER (5)

*Surface resistivity or conductance* (1, 5, 7, 8, 23). Some hygroscopic materials exhibit a large change in electrical resistance with a relatively small change in moisture content. Such materials attract or release moisture from the surrounding atmosphere in proportion to the relative humidity. Measurement of their electrical resistance can then be used to determine relative humidity. Resistance sensors will usually take the form of a thin film of hygroscopic material over a pair of intermeshing electrodes on a plastic base.

An early electrical resistance hygrometer was developed by Dr. F.W. Dunmore at the National Bureau of Standards and is known as the Dunmore cell. The sensor depends on the resistance change of a bifilar element embedded in a lithium chloride film on a plastic base. The resistance of the film varies with humidity and governs the current that will pass across it between the wires. Sensitivity to relative humidity requires that bifilar element spacing and/or film resistance properties be varied for specific humidity ranges. The sensitivity results in several resistance elements being required to cover the complete relative humidity range.

Materials other than lithium chloride have also been used for the hygroscopic resistance element. The Pope cell uses a polystyrene substrate; other electrical sensors are fabricated of electrolytic material, ceramics, and polyvinyl chloride. A carbon strip element is used in Radio Sonde devices sent aloft to measure pressure, temperature, and humidity for weather forecasting purposes.

Absolute Moisture Measurement: *Gravimetric* (1, 2, 3, 5, 6). The simplest and most direct way to measure water content of a solid, liquid, or a gas is to compare the weights of a sample before and after moisture has been removed. Heating is a common method for removing moisture from solids while a gas is passed through a drying agent. The latter technique is considered to be the current state of the art for precision humidity instrumentation and is used for standardization and calibration. The gravimetric hygrometer developed and maintained by NBS yields an absolute humidity measurement. A measured quantity of test atmosphere is passed through a tube containing a drying agent, and the increase in weight of the drying agent is measured.

While the gravimetric method of moisture measurement would appear simple and straight-forward, experience shows that a great deal of technique is involved (13). Difficulties include the driving off of volatile liquids other than water; confusion between bound water and free water; the history of the sample between the time it is taken and the time it is analyzed (13); and the precision of the weighing balance. Further, the gravimetric method is quite unwieldy for low-humidity samples, which require up to 30 hours per calibration point, and, of course, its output cannot be made continuous for transmission. With proper technique, however, the gravimetric hygrometer is more accurate than other systems, as seen in Fig. 4-13, which shows a graph of uncertainty for several devices.

 b. Cannot be used on-line
 c. Very time consuming for low moisture content
B. Hygroscopic sensors
 1. Capacitive sensors (see Figs. 4-15 and 4-16)
  a. Theory of operation
   (1) Capacitor is formed by two electrically conductive elements separated by a less conductive medium called a dielectric
   (2) The capacitance varies with the dielectric constant of the intervening medium
   (3) Dielectric is composed of material that shows marked change in dielectric constant with changes in moisture content
   (4) Measurement of capacitance provides indication of moisture content of material
   (5) Measurement is read out by a capacitive bridge-type circuit
  b. Range and accuracy
   (1) Range of 0 to 15% moisture
   (2) Accuracy of 0.2 to 1%
  c. Advantages
   (1) Continuous readout
   (2) Simplicity
  d. Disadvantages
   (1) Temperature sensitive
   (2) Dielectric constant sensitive to density, packing, particle size
   (3) Hysteresis
C. Electrolytic hygrometers (see Figs. 4-17 and 4-18)
 1. Theory of operation
  a. Passage of a direct current through a moisture-bearing sample causes decomposition into the constituent elements of hydrogen and oxygen
  b. The current necessary to electrolyze the moisture varies as the amount of moisture present
 2. Range and accuracy
  a. Range of 0 to 1000 ppm of moisture
  b. Accuracy of ±5%
 3. Advantages
  a. Very low moisture contents can be measured
 4. Disadvantages
  a. Sensitive to hydrogen and oxygen in sample
  b. Difficulty of zeroing instrument

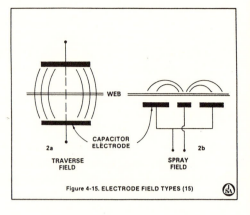

Figure 4-15. ELECTRODE FIELD TYPES (15)

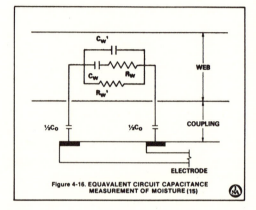

Figure 4-16. EQUAVALENT CIRCUIT CAPACITANCE MEASUREMENT OF MOISTURE (15)

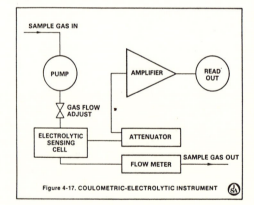

Figure 4-17. COULOMETRIC-ELECTROLYTIC INSTRUMENT

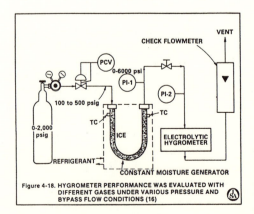

Figure 4-18. HYGROMETER PERFORMANCE WAS EVALUATED WITH DIFFERENT GASES UNDER VARIOUS PRESSURE AND BYPASS FLOW CONDITIONS (16)

*Hygroscopic instruments* (1, 2, 3). A number of moisture sensors depend on the change in electrical properties of a material as they absorb water from the surroundings. The electrical properties that are varied are usually resistance or impedance. Capacitive sensors use the change in the dielectric constant of a material due to the presence of water or water vapor. A dessicant material is used to isolate the plates of a capacitor. Water absorption changes the capacitance of the isolating material.

Measurements are made using an ac bridge that reads directly.

The same technique is used to measure the amount of water in a solid. In this case, the solid is passed between parallel capacitor plates. This technique is widely used in the paper industry (15). The dielectric constant of paper is about 2; that of water about 81; thus the dielectric properties of paper change when moisture is absorbed in the fiber structure. The moisture-measuring element is made up to two electrode elements that span the paper sheet. One electrode is the reference electrode, and the second is a sensing electrode, A frequency generator supplies alternating current to both electrodes. The plates of the capacitor made up of the reference electrodes are subjected to a constant dielectric condition, while the sensing plates transmit a varying ac signal because of the varying dielectric constant. The signals are amplified and transmitted to an ac bridge circuit, which is unbalanced by the difference in reference and detector signals.

*Electrolytic hygrometers* (1, 2, 3, 14). These hygrometers are also called coulometric hygrometers because they measure moisture in a gas by measuring the current (coulombs per unit time) needed to completely electrolyze the moisture. A sample is passed at a controlled rate through a cell containing a dessicant, such as phosphorus pentoxide, which completely absorbs the moisture in the sample. A dc voltage is applied to platinum electrodes within the cell. The amount of current required to dissociate the absorbed water vapor into hydrogen and oxygen is measured, amplified, and transmitted as an indication of moisture content.

Limitations of the electrolytic hydrometer include errors due to the recombination of hydrogen and oxygen into water, and the absorption of the water by materials in contact with the sample gas (16). For these reasons, many modifications of the instruments exist. The original configuration had platinum wires inside a Teflon® capillary. Then the wires were wound on the outside of a Teflon tube to eliminate clogging the capillary, and the gas was passed through the annular space outside the tube. The aforementioned recombination effect is especially severe with gases containing hydrogen or oxygen, and the platinum wires have been replaced with rhodium in this service to minimize this effect. Another variation is to measure the moisture at two different flow rates and to determine the true reading by difference to minimize the effect of meter zero, which is very difficult to determine. The measurement is very sensitive to the density, pressure, and flow rate of the sample.

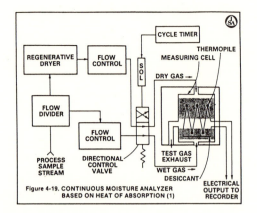

Figure 4-19. CONTINUOUS MOISTURE ANALYZER BASED ON HEAT OF ABSORPTION (1)

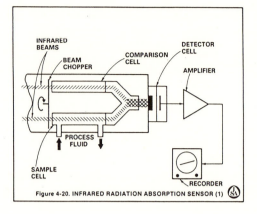

Figure 4-20. INFRARED RADIATION ABSORPTION SENSOR (1)

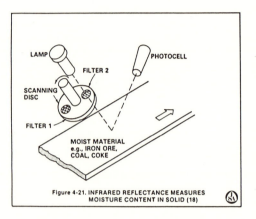

Figure 4-21. INFRARED REFLECTANCE MEASURES MOISTURE CONTENT IN SOLID (18)

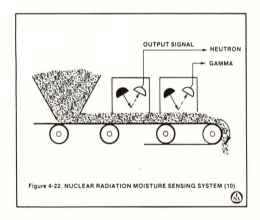

Figure 4-22. NUCLEAR RADIATION MOISTURE SENSING SYSTEM (10)

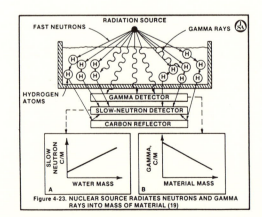

Figure 4-23. NUCLEAR SOURCE RADIATES NEUTRONS AND GAMMA RAYS INTO MASS OF MATERIAL (19)

D. Heat of adsorption (see Fig. 4-19)
  1. Theory of operation
    a. The condensing of moisture onto the surface of a dessicant will heat the dessicant; evaporation from the surface will cool the dessicant
    b. The heat energy exchanged by the adsorption and desorbtion causes the temperature of the adsorbent to change
    c. Temperature change varies with the concentration of moisture in the gas stream
  2. Range and accuracy
    a. Ranges of 0 to 10 ppm to 0 to 5000 ppm
    b. Accuracy ±1%
E. Infrared radiation absorption (see Figs. 4-20 and 4-21)
  1. Theory of operation
    a. Water molecules can be excited by a resonant absorption phenomenon in several frequency bands in the infrared spectrum
    b. Radiant energy of a resonant frequency is absorbed by the water molecule
    c. Radiation of nonresonant frequencies passes through the water molecule unattenuated
    d. The amount of energy absorbed in the water absorption band is a measure of the moisture content of the sample
  2. Range and accuracy
    a. Range of 0 to 300 ppm
    b. Accuracy ±1%
  3. Advantages
    a. Used with solids, liquids, gases
    b. Rapid response
  4. Disadvantages
    a. Complicated and expensive
    b. Frequent calibration required
    c. Water induced absorption masked by other sample constituents
F. Nuclear radiation absorption (see Figs. 4-22 and 4-23)
  1. Theory of operation

*Heat of Adsorption** (1, 2, 3). This process is used in the thermal hygrometer, in which the heat evolved in the adsorption and desorption of moisture on a dessicant material is used as a measure of humidity. The adsorption and desorption processes release and absorb heat, respectively, and the temperature change is indicative of the moisture content of the sample.

In practice, the monitored sample and a reference gas are alternatively cycled through a column of adsorbent material, and thermopiles or thermistors embedded in the adsorbing material are alternatively exposed to dry gas and sample gas.

*Nuclear Magnetic Resonance* (17). NMR analyzers, formerly described in the section on "Flow Measurement," can be used to determine the moisture content of materials. When the molecules in the flowing stream are irradiated with radio frequency excitation signals, they are caused to resonate. The relaxation of the induced magnetic field is monitored as the molecules return to ground state. Components in the flowing stream can be identified since different molecules produce magnetic echoes at distinct times.

For example, in some solutions the signals from other components will decay much faster than the signals generated by the water. The observed echo is due only to the water molecules and can be calibrated to the relative water content by the ratio of the magnitude of the echo signal to the orginal pulse height.

*Infrared Radiation Absorption* (1, 2, 3, 5, 7, 18). Infrared spectrometers detect moisture in a liquid sample by measuring the energy absorbed in the water vapor band (between wavelengths of 0.1 and 1.9 microns). The spectrometer is a complex and expensive instrument. It consists of an energy source, a radiant energy detector, an optical system for isolating the wavelengths of interest, and a sensor to measure energy attenuation in the optical path.

Infrared reflectance has been used to measure the moisture content of a solid, such as iron ore (18). The ore is a strong absorber of IR wavelengths because of its conductance and black color. A scanning disc, carrying two interference filters, causes two infrared wavelengths to fall on the moist ore. The reflected light is measured by lead-sulfide photoelectric cells. The ratio of the reflected radiation at these two wavelengths is a measure of the percentage moisture content of the material. This method can be applied to any dark material and is unaffected by packing density, changing composition, or temperature.

*Nuclear radiation absorption* (1, 2, 19) depends on the retarding effect of the hydrogen atoms in water on the speed of the neutrons to sense the moisture. Since the flow of neutrons from source to detector is also retarded by the mass of the material, simultaneous density measurements must be made and used as compensation. The sensor will typically consist of a high-energy neutron source and a detector to measure moisture, plus a gamma ray source and a detector for the density measurement.

Nuclear radiation has been successfully applied to the determination of moisture in raw materials on a conveyor (19). A nuclear source transmits both fast neutrons and gamma rays through the material. In passing through the material the fast neutrons collide with the hydrogen atoms like billiard balls, slowing enough to be sensed by the slow neutron detector. The detector generates voltage pulses. Since water consists of both hydrogen and oxygen atoms, the counting rate varies directly with the amount of moisture in the gaging zone.

---

**Adsorption* is the adherence of the molecules of a fluid to the surface of another substance, while *absorption* is the penetration of the fluid into the inner structure of the substance.

a. The hydrogen in water absorbs fast neutrons at a rate appreciably greater than any other element

b. Measurement of absorption of neutron energy by a sample is a measure of moisture content per unit of volume

c. Since most samples exhibit density variations, the neutron absorption must be corrected for density

2. Range and accuracy
   a. Range of 2 to 80% moisture
   b. Accuracy ±0.25%

3. Advantages
   a. Accuracy
   b. Indicates moisture/unit weight

4. Disadvantages
   a. Sensitive to other bound hydrogen in sample
   b. Complex and expensive

G. Microwave hygrometers (see Fig. 4-24)
   1. Theory of operation
      a. Metallic electrodes are placed in or near the sample, and radio frequency voltage applied to them
      b. Water molecules strongly absorb the microwaves from the radio frequency excitation
      c. The absorption results in a change in the amplitude and phase of the microwaves
      d. The measured change in amplitude and phase varies with the moisture content

   2. Range and accuracy
      a. Range is from 0 to 80% moisture
      b. Accuracy is from ±0.1% to ±1%, depending on the sample composition

   3. Advantages
      a. Applicable to gases, solids, liquids
      b. High selectivity to moisture
      c. Fast response

   4. Disadvantages
      a. Expensive
      b. Sample cannot be electrically conductive
      c. Absorption varies with temperature and density

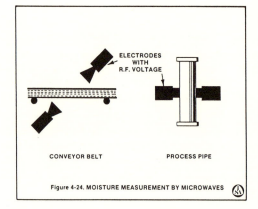

Figure 4-24. MOISTURE MEASUREMENT BY MICROWAVES

On the other hand, the gamma ray passing through the material interacts with the electrons of all the atoms of the material. This interaction causes gamma ray scattering and absorption; and the amount of gamma radiation reaching the detector is inversely related to the amount of material between source and detector. The signals from the two detectors, after electronic conditioning, are divided, resulting in an indicated measurement of moisture per unit weight.

*Microwave hygrometers* (1,2,20,21,22). Radio frequency power is applied through metallic electrodes in the microwave moisture sensor. The microwaves are transmitted through the sample under test. Rotational relaxation, a spectographic phenomenon, causes water to strongly absorb microwaves. The absorption causes a change in ɩhe amplitude and phase of the waves. Moisture analysis is by the electronic measurement of wave amplitude and phase.

The important factors in the moisture measurement are the energy absorbed by the water relative to the other components present and the frequency of maximum absorption. Water is 1000 times as absorptive as most other materials. Thus it is nearly independent of base material or sample composition. Local absorption peaks about $2.2 \times 10^{10}$ cps. The microwave technique can be applied to solids, liquids, or gases. Errors are introduced by changing temperature and a change in the ratio of free-to-bound water.

# REFERENCES

1. Stalhuth. W.E., "Moisture Measurement and Control," *Automation,* November, 1964, p. 84.
2. Anon., "Survey of Humidity Instrumentation," *Instruments and Control Systems,* January, 1972, p. 75.
3. Martin, S., "Hygrometers," *Control* (British), July, 1965, p. 371.
4. McGill, R.J., "Mositure Control-Principles, Practice, and Problems," *Measurements and Control* (British), October, 1969, p. 177.
5. Cole, K.M. and J.A., Reger, "Humidity Calibration Techniques," *Instruments and Control Systems,* January, 1970, p. 77.
6. Nelson, R.C., "Moisture and Humidity," *Instruments and Control Systems,* August, 1964, p. 82.
7. Amdur, E.J., "Humidity Sensors," *Instruments and Control Systems,* June, 1963, p. 93.
8. Whitehaus, G., "Linearizing Relative Humidity Measurements," *Instruments and Control Systems,* September, 1972, p. 72.
9. Zwart, H.C., "Errors in Humidity Elements," *Instruments and Control Systems,* August, 1966, p. 95.
10. Hygrodynamics Technical Bulletin No.4, Hygrodynamics, Inc., 949 Selim Road, Silver Spring, Md.
11. U.S. Department of Commerce, Weather Bureau Psychrometric Tables, WB235.
12. Hygrodynamics Technical Bulletin No. 1, Hygrodynamics, Inc., 949 Selim Road, Silver Spring, Md.
13. Boonton Polytechnic Co., "General Remarks on the Calibration of Moisture Analyzers by Weight Loss Methods," Boonton Polytechnic Co., 14 Union St., Rockaway, N.J., April, 1964.
14. Long, D.O., "Low Humidity Test Measurments," *ISA Transactions,* Vol. 8, No. 1, p. 72.
15. Slocomb, J.W., J.R. Parkinson, and D.D. Wilma, "Capacitive Moisture Measurement," *Instruments and Control Systems,* March, 1969, p. 131.
16. Belkin, H.M., D. Jhaveri, and M.M. Kapani, "Getting the Most Out of Electrolytic Hygrometers," *Instrumentation Technology,* July, 1970, p. 57.
17. DuPont Co., "Pulsed NMR Applications Lab Report," Instrument Products Division, Wilmington, Del.
18. Anon., "Infrared Reflectance Measures Iron Ore Moisture," *Control Engineering,* June, 1973, p. 23.
19. Reim, T.E., "Advanced Nuclear Gage Controls Process Moisture," *Instrumentation Technology,* July, 1967, p. 47.
20. Busker, L.H., "Microwave Moisture Measurement," *Instruments and Control Systems,* December, 1968, p. 89.
21. Bulletin 700B, Model 700 Microwave Moisture Gage, Microwave Instruments Co., 3111 Second Ave., Corona DelMar, Calif. 92625.
22. Miller, P.S., and J.J. Jones, "Moisture Deterministation by Means of Radio Frequency Power Absorption," Boonton Polytechnic Co., Rockaway, N.J.
23. Cortina, V., and R. Budnauito, "Precision Humidity Analysis," Proceedings of 1979 ISA Symposium, Newark, DE, June, 1979, p. 166.
24. Hayes, S., "Chilled Mirror Hygrometry," *Measurements and Control,* February, 1981, p. 115.
25. Cucchiara, O., "Aluminum Oxide Hygrometry in Gases," Proceedings of 1979 ISA Symposium, Newark, DE, June, 1979, p. 182.
26. Black, S.D., "Piezoelectric and Electrolytic Methods of Moisture Measurement in Gases," Proceedings of 1979 ISA Symposium, Newark, DE, June, 1979, p. 187.
27. Weisen, R., "Detection of Moisture in a High Temperature Process Gas Stream," Proceedings of 1979 ISA Symposium, Newark, DE, June, 1979, p. 191.
28. Weiderhold, P.R., "Which Humidity Sensor?" *Instruments and Control Systems,* June, 1978, p. 31.
29. Bailey, S.J., "Moisture Sensors in 1980; On-Line Roles Increases," *Control Engineering,* September, 1980, p. 112.
30. Yang, L.C., "Developments in Moisture Sensors," *Measurments and Control,* February, 1981, p. 98.
31. Anon., "Moisture Analyzer Helps Extend Period Between Dryer Column Recharging," *Control Engineering,* August, 1980, p. 31.

# 5
# PHYSICAL AND CHEMICAL MEASUREMENTS

## INTRODUCTION

Three of the four basic measurements (flow, pressure, level, and temperature) formerly discussed have to do with the amount or quantity of a material. Flow describes its rate of movement, while accumulation is measured by pressure or level. Only temperature has to do with the condition or quality of the material. In some cases, temperature completely describes the condition of the material, as with the temperature of boiling and freezing water under ambient conditions.

Temperature alone is inadequate to describe the condition of complex chemicals, or mixtures, how-ever. For example, it has been pointed out that plant effluent (waste discharge) is fully described by four measurements — pH, conductivity, dissolved oxygen, and temperature (1). Thus additional measurements are required. Some of these have already been discussed — density and moisture measurement, for example. Instruments have become available, in many cases quite recently, for the measurement of many other physical and chemical properties. Most of these sensors have been adapted from laboratory instruments, and the early adaptations were too fragile for use in an operating environment. Therefore, continuous maintenance was required. However, rugged versions have now been developed, and these sensors can be expected to operate reliably for extended periods with only minimum protection from extreme service that would cause failure.

Turbidity is defined as an optical appearance property of liquids caused by the presence of suspended particles. The particles cause a scattering of the light energy passing through the liquid, and the turbidiy is influenced by the concentration, size, shape, and optical properties of the particles in addition to the optical properties of the fluid (3).

The theory of the scattering of light by particles was developed late in the nineteenth century. A theoretical understanding of the light-scattering phenomena by particles is well established, but the general theories are so complex that they require computers for analytical solutions. However, seldom is this complexity required in the measurement of the turbidity in industrial applications (3).

Viscosity is a fluid's resistance to flow or to a change in form. It is a measurement of the ease with which the particles that make up a fluid can yield to a force and change their relative positions. Viscosity reflects the combined effect of the molecular interchange of momentum and the molecular cohesive forces in gases and liquids (6). The molecules of all substances are in a constant state of agitation. The agitation causes a continual interchange of momentum between molecules as they collide. Further, the molecules move back and forth between adjacent layers of fluid. If adjacent layers are moving at different velocities, the constant interchange of molecules and momentum creates resistance to the relative motion of the layers. Further, in liquids, the molecules are cohesive, which causes them to bind in position and resist a change in shape. The cohesiveness and the interchange of molecules both create resistance to distortion, which is called viscosity.

In considering the laminar flow of a viscous fluid, Newton assumed a constant velocity gradient. That

### Table 5-1
### Viscosity Terminology and Equations (7)

**Stress** — Force/Area — F/A

**Velocity Gradient (Shear)** — Rate of change of liquid velocity across the stream — V/L for linear velocity profile, dV/dL for nonlinear velocity profile. Units are $V/L = ft/sec/ft = sec^{-1}$.

**Absolute (dynamic) viscosity** $\eta$— Constant of proportionlaly between applied stress and resulting shear velocity (Newton's Hypothesis):

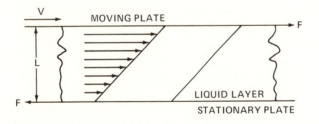

$$\frac{F}{A} = \frac{\eta V}{L} \quad ;\eta = \frac{F/A}{V/L}$$

**Poise** $(\eta)$ — Unit of dynamic or absolute viscosity — 1 dyne-sec/cm².

**Kinematic Viscosity** $\nu$ — Dynamic viscosity/density $= \eta/\rho$.

**Stoke** — Unit of kinematic viscosity $(\nu)$ = 1 cm/sec.

**Hagen-Poiseuille Law** (flow through a capillary) —

$$Q = \frac{\pi R^4}{8 \eta L} (P_1 - P_2)$$

**Saybolt Viscometer (Universal, Furol)** — Measures time for given volume of fluid to flow through standard orifice; units are seconds.

**Fluidity** — Reciprocal of absolute viscosity; unit in the cgs system is the rhe, which equals 1/poise.

**Specific Viscosity** — Ratio of absolute viscosity of a fluid to that of a standard fluid, usually water, both at same temperature.

**Relative Viscosity** — Ratio of absolute viscosity of a fluid at any temperature to that of water at 20°C (68°F). As water at this temperature has an $\eta$ of 1.002 cp, the relative viscosity of a fluid equals approximately its absolute viscosity in cp. As density of water is 1, kinematic viscosity of water equals 1.002 cs at 20°C.

**Apparent Viscosity** — Viscosity of a non-Newtonian fluid under given conditions. Same as consistency.

**Consistency** — Resistance of a substance to deformation. It is the same as viscosity for a Newtonian fluid and the same as apparent viscosity for a non-Newtonian fluid.

**Saybolt Universal Seconds (SUS)** — Time units referring to the Saybolt Viscometer.

**Saybolt Furol Seconds (SFS)** — Time units referring to the Saybolt viscometer with a Furol capillary, which is larger than a Universal capillary.

**Shear Viscometer** — Viscometer that measures viscosity of a non-Newtonian fluid at several different shear rates. Viscosity is extrapolated to zero shear rate by connecting the measured points and extending curve to zero shear rate.

is, the change in velocity between adjacent infinitesimally thin layers of fluid is constant. Many common fluids exhibit this characteristic and are designated as Newtonian fluids. However, viscosity measurement is complicated by the fact that the viscosity in some fluids varies with the velocity of flow (7). Catsup is a common example — it is difficult to start flowing (high viscosity) but flows very easily when started (low viscosity). Such fluids are designated as non-Newtonian, or pseudoplastic (8).

The term consistency is shown in Table 5-1 as the equivalent of viscosity in referring to a fluid that is a slurry, such as the pulp slurry used in the manufacture of paper (17).

Three basic forms of instruments are in use for the precise viscometry of Newtonian fluids (8). Dynamic viscosity is measured by applying a shear force to the fluid and measuring the resulting velocity. Sliding plate or rotating cylinder sensors are used to measure dynamic viscosity. A second basic type of viscometer measures the viscosity by timing the flow of a given volume of liquid. This type measures "kinematic" viscosity because the word kinematic means pertaining to flow. Finally, the falling-ball, or rolling-ball, viscometer requires the measurement of the time that it takes for a meter ball to fall or roll from one end of the tube to the other. The time is then a measure of viscosity. Table 5-2 shows a listing of currently available viscometers using these basic techniques.

Table 5-2
Viscometer Characteristics (9)

| Type | Viscosity range (cp) | Temp range (F) | Cost ($) | Accuracy (%) |
|---|---|---|---|---|
| Laboratory | | | | |
| Capillary | 0.2-120,000 | –100 to 300 | 10-2,700 | 0.35 |
| Efflux cup | 1-1200 | 60 to 250 | 15-30 | 0.1 |
| Falling ball | 0.01-160,000 | –30 to 300 | 350 | 0.1 |
| Bubble | 0.5-125,000 ctks | 77 | 5-75 | 2.0 |
| Rotational | $10^{-4}$-$10^8$ | –10 to 2500 | 45-24,000 | 0.5 |
| Plastometers | 0-200 Mooney 0-200 MI 0-100 CIL | to 570 | 2,000-20,000 | 1.0 |
| Industrial | | | | |
| Rotational | 0-$10^6$ | 160 case 2500 for fluid | 1,000-2,5000 | 1.0 |
| Falling piston | 0.1-$10^6$ | to 650 | 1,100-2,500 | 1.0 |
| Float | 0.5-$10^4$ | 450 | 300-3,500 | 2.0 |
| Continuous capillary | 0-$10^6$ | to 900 | 1,500-7,000 | 0.5 |
| Ultrasonic | 0-50,000 ctks | –190 to 650 | 1,300-2,500 | 2.0 |
| Vibrating reed | 0.1-$10^5$ | –150 to 300 | 1,500 | 1.0 |

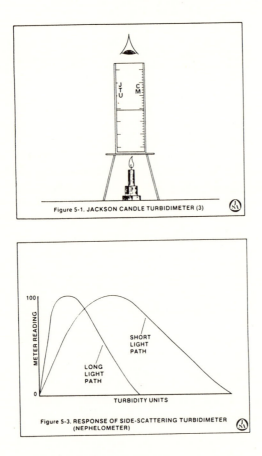

Figure 5-1. JACKSON CANDLE TURBIDIMETER (3)

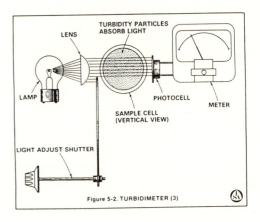

Figure 5-2. TURBIDIMETER (3)

Figure 5-3. RESPONSE OF SIDE-SCATTERING TURBIDIMETER (NEPHELOMETER)

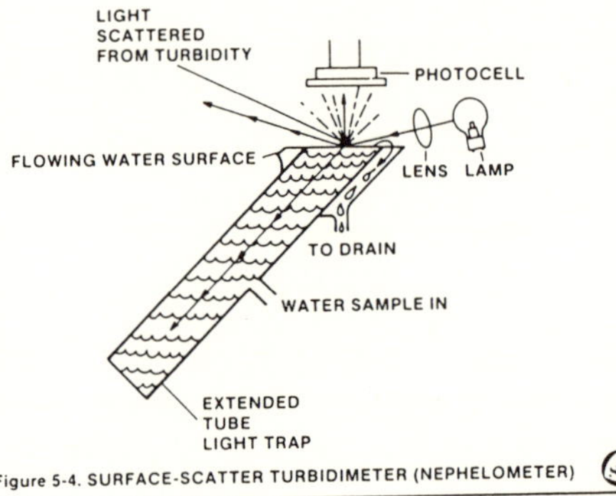

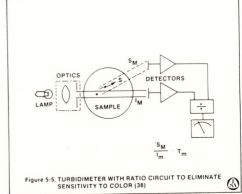

Figure 5-4. SURFACE-SCATTER TURBIDIMETER (NEPHELOMETER)

I. Turbidity and turbidimeters (see Fig. 5-2)
  A. Principle of operation
    1. A light is directed through a liquid sample held in a transparent chamber
    2. Solid particles in the liquid will absorb and scatter the light
    3. Light energy that passes through the chamber is measured and related to the cloudiness of the sample
  B. Difficulties of measurement
    1. Light scattering varies with particle shape, size, refractive index, and frequency spectrum of light source
    2. Little correlation among measurements of various turbidimeters
    3. Measurement sensitive to scratches, imperfections, dirt, and film on sample chamber
  C. Nephelometer (see Fig. 5-4)
    1. Highly sensitive for measuring small turbidities
    2. Zero signal at zero turbidity
    3. Output signal increases with increasing turbidity
    4. Output signal linear with turbidity in low ranges
  D. Absorptometer (see Fig. 5-5)
    1. Not sensitive for small turbidities

Figure 5-5. TURBIDIMETER WITH RATIO CIRCUIT TO ELIMINATE SENSITIVITY TO COLOR (38)

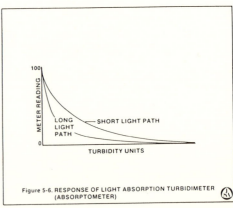

Figure 5-6. RESPONSE OF LIGHT ABSORPTION TURBIDIMETER (ABSORPTOMETER)

Turbidity is an optical property of a liquid. The Jackson candle turbidimeter, shown in Fig. 5-1, and the Jackson Turbidity Unit are the standard instrument and unit of turbidity measurement, respectively. The Jackson turbidimeter is a special candle and a flat-bottomed glass tube graduated in Jackson Turbidity Units. The turbid sample is poured into the tube as the image of the candle is observed from the top. When the image disappears into a uniform glow, a reading on the depth of the sample is taken (4).

The Jackson turbidimeter compares the strength of the transmitted light with that scattered or reflected. The yellow-red candle flame is at the long wavelength end of the visible spectrum and is not effectively absorbed or scattered by very fine particles. Thus the candle flame will not disappear, and the Jackson turbidimeter cannot be used for such particles. Further, black particles absorb all of the light, and the liquid will become dark before enough sample is poured into the tube to reach an image extinction point (4).

These difficulties are partially circumvented in modern instruments using incandescent light sources, which provide a wider light spectrum, or by using comparison rather than extinction techniques for measurement. The nephelometer is a turbidimeter that measures the right-angle light scatter by a fluid and, thus, is highly sensitive to low turbidities. The absorptometer, on the other hand, is a turbidimeter that measures light passing through the sample. It is insensitive to small turbidities but has no upper limit on the turbidity range (4). The characteristics of the nephelometer and absorptometer follow.

*Nephelometer.* Figure 5-3 shows the response of a nephelometer to be zero at zero turbidity and that the signal is proportional to turbidty up to a value where it is blinded by the opacity of the process liquid (36).

Sensitivity and the span of measurement can be changed by choice of light path, light brightness, and photocell sensitivity. A second detector used in a ratioing circuit can be used to eliminate sensitivity to color (37, 38).

*Absorptometer.* Figure 5-6 shows the response of an absorptometer to be maximum at zero turbidity, and to have an output signal that decreases with increasing turbidity (36).

The output is less sensitive to changes in turbidity than the nephelometer and is also sensitive to color.

2. Maximum signal at zero turbidity
3. Output signal decreases with increasing turbidity
4. Color sensitive
E. Turbidimeter characteristics
   1. Accuracy of ±5% after calibration
   2. Repeatability ±1%
   3. Temperature range 40 to 250°F
   4. Maximum pressure 100 psi

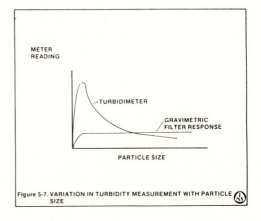

Figure 5-7. VARIATION IN TURBIDITY MEASUREMENT WITH PARTICLE SIZE

II. Viscosity and viscometers
   A. Capillary viscometer (see Fig. 5-8)
      1. Theory of operation
         a. Fluid is forced through capillary at constant flow rate
         b. Capillary is maintained at constant temperature
         c. Pressure drop across capillary is proportional to viscosity
      2. Characteristics
         a. Viscosity range 0 to 2500 cp
         b. Required temperature regulation to ±0.005°F
         c. Repeatability ±0.5%
         d. Response time, 0.8 minute dead time and 0.7 minute time constant
         e. Continuous operation
      3. Limitation
         a. Not applicable to non-Newtonian fluids

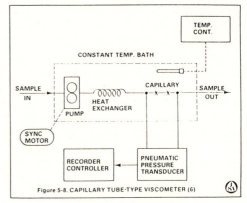

Figure 5-8. CAPILLARY TUBE-TYPE VISCOMETER (6)

   B. Falling ball viscometer (see Fig. 5-9)
      1. Theory of operation
         a. High precision ball falls through liquid contained in high-precision glass tube
         b. Densities of ball and liquid are known
         c. Time for ball to fall length of tube is measured
         d. Viscosity is proportional to difference in density and to time
      2. Characteristics
         a. Range 0.25 to 300 cp
         b. Repeatability of ±0.5%
         c. Discrete samples

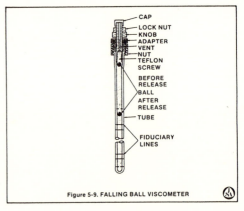

Figure 5-9. FALLING BALL VISCOMETER

   C. Falling piston viscometer (see Fig. 5-10)
      1. Theory of operation
         a. Sample solenoid valve opens to raise the piston
         b. Piston travels up in 3 seconds
         c. Sample solenoid valve closes when piston reaches top of travel
         d. Time for piston to fall is a measure of viscosity
         e. Time starts when piston begins to fall, ends

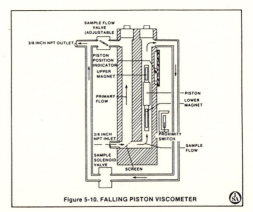

Figure 5-10. FALLING PISTON VISCOMETER

The turbidity measurement is unique to the type of turbidimeter used, and thus calibration relating to the quantity of suspended matter can be made only under very restricted conditions (36, 37, 38, 39). Two different types of turbidimeters standardized to the same turbidity standard can show a variation of as much as 500 percent when applied to the same sample of unknown turbidity (36). The variation in turbidity reading with particle size is shown by Fig. 5-7.

Thus even comparisons restricted to turbidity measurements made on the same turbidimeter using samples containing the same process material will usually not exhibit a linear quantitative relationship between suspended matter and turbidity value (36).

The modern turbidimeter uses a voltage-regulated light source mounted on one side of a sample chamber. The light projects a beam of energy through the liquid sample. As the energy is projected through the fluid, a part of the radiant energy is deflected or absorbed by the particulate matter. The radiant energy reaching the far side of the sample chamber is detected by a photometric detector, usually a photocell or a bolometer, which produces an electrical signal that is related to the energy detected. The energy-detection device is made into one arm of a Wheatstone bridge measuring circuit. Measurement is then made and transmitted by a null-balance amplifier, which continuously positions a servomotor and measuring slidewire to maintain the circuit in balance (1, 5).

*Capillary Viscometer.* The following formula, known as the Hagen-Poiseuille law, was discussed under "Capillary Flowmeters" in the section on flow:

$$Q = K \frac{\Delta p}{\eta L}$$

where

$Q$   =   volumetric flow rate

$K$   =   constant

$\Delta p$   =   pressure drop

$\eta$   =   viscosity

$L$   =   tubing length

The capillary device is one in which flow rate through a small tube of known diameter under a hydrostatic head is measured. When used as a flowmeter, the differential pressure is measured and related to flow rate. However, viscosity must be maintained constant. But the foregoing equation can also be written:

$$\eta = K \frac{\Delta p}{Q L}$$

Now the viscosity can be determined by measuring the differential pressure and maintaining the flow constant, The capillary metering technique was developed in 1840 by Dr. Poiseuille, a French physician, who was interested in the behavior of blood as it flows through the veins in the body. The poise, a unit of dynamic viscosity, is named for Dr. Poiseuille.

*Falling-Ball Viscometer.* The terminal velocity of the descent of a sphere is measured in the falling-ball viscometer. The density and diameter of the sphere, or ball, are known, as is the density of the fluid under test. The ball is forced down by the difference in its density and that of the fluid. The measurement is sensitive to fluid density, since an increase in density will cause the ball to fall more slowly, indicating a higher viscosity. Thus the falling ball viscometer measures absolute viscosity but requires a density correction (7, 8).

Neither the capillary nor the falling ball viscometer is satisfactory for measurements on non-Newtonian fluids (8). These devices produce a shear rate in the fluid, which varies continuously from zero (at the center of the capillary tube or at a distance from the ball) to a maximum value (at the wall of the capillary or the surface of the ball). If the velocity gradient is linear, as with a Newtonian fluid, the viscosity measurement is independent of the size of the ball or the diameter of the capillary. The nonlinear velocity gradient that characterizes a non-Newtonian fluid will result in a different viscosity reading for every change in parameters such as ball density or capillary diameter.

when magnet on piston actuates proximity switch
2. Characteristics
    a. Maximum conditions, 500 psig and 300°F
    b. Range — 1 to 6500 cp
    c. Newtonian and non-Newtonian liquids
    d. Continuous sampled output

D. Sliding plate viscometer (see Fig. 5-11)
1. Theory of operation
    a. Fluid sample held between two parallel plates
    b. Plate A is fixed; plate B is connected to loading weighbeam
    c. Weight on weighbeam causes plate B to move
    d. Servomotor follows movement
    e. Output signal is a function of time
2. Characteristics
    a. Viscosity range of 1000 to 100 billion poise
    b. Film thickness of 10 to 200 microns
    c. Loads of from 0.1 gm to .10 kg
    d. Constant temperature bath required
    e. Non-Newtonian fluids can be accommodated by taking several readings with different weights on the weighbeam

E. Rotational viscometer (see Figs. 5-12 and 5-13)
1. Theory of operation
    a. Coaxial cylinders or parallel disc is rotated at constant speed
    b. Viscous drag incurs torque on rotating spindle
    c. Torque to cause spindle to rotate is a measure of viscosity
2. Characteristics
    a. Viscosity range of 500 to 5000 poise
    b. Repeatability ±2.5%
    c. Continuous output signal
    d. Multiple-radius spindles can be used with non-Newtonian fluids

F. Float-type viscometer (see Fig. 5-14)
1. Theory of operation
    a. Upward flow causes differential pressure float to position such that differential pressure across viscometer is constant
    b. Constant differential pressure causes constant flow around float
    c. Float has large viscous drag area and is carried upward by viscous drag of fluid
    d. Float position is a measure of viscosity
2. Characteristics
    a. Viscosity range of 0.5 to 550 centistokes

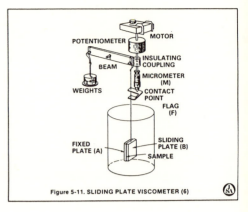

Figure 5-11. SLIDING PLATE VISCOMETER (6)

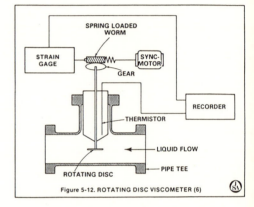

Figure 5-12. ROTATING DISC VISCOMETER (6)

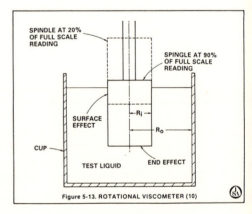

Figure 5-13. ROTATIONAL VISCOMETER (10)

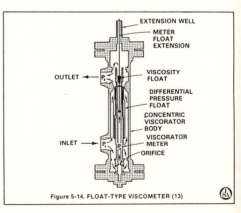

Figure 5-14. FLOAT-TYPE VISCOMETER (13)

*Sliding Plate Viscometer.* This viscometer is based on the principle of the shearing of a sample of fluid between two parallel plates under the action of a constant shearing stress. The plates are of Pyrex® glass, and can be used with plastics and other viscous materials that adhere to the plate material. The plates are only 100 microns (micrometers) apart, and thus the device is sometimes called a microviscometer (6).

The plates are oriented in a vertical position during operation. One is clamped in place, and the other is attached to a loading weighbeam. A loading weight is attached to the opposite end of the weighbeam from the sliding plates. A servomechanism follows the motion of the weighbeam, and the movement of the plate with time is printed out on a millivolt recorder.

*Coaxial Cylinder Viscometers.* These viscometers are similar to the sliding plate type except that the plates are cylindrical, and one is fitted inside the other. The inner cylindrical spindle rotates in the concentric outer cup, and for this reason the sensor is sometimes called a *rotational viscometer.* There are two basic designs of rotational viscometers: (a) the inner cylinder rotates at constant rpm, and the torque on the stationary outer cylinder is measured; and (b) the inner cylinder is rotated at the constant torque, and the outer cylinder merely acts as a container to hold the fluid (6, 8).

The torque required to rotate a disc at a constant low speed in a fluid is directly related to viscosity. The width of the annulus between the two cylinders is critical to the measurement. If the width is large (0.1 inch) so that solid particles in the fluid can pass through the viscometer, an upper limit is imposed on the shear rates that can be obtained before the flow regime changes from laminar to turbulent. One viscometer has a gap width of 0.0002 inch, which permits the measurement of very high viscosities, but is susceptible to plugging by very small solid particles (8, 10)

Finally, the mechanical work that must be dissipated in the measurement of very high viscosities leads to the accumulation of heat in the fluid, and an increasing temperature causes a change in viscosity. Thus rotational viscometers used in high-viscosity service require a temperature bath.

The rotational viscometer can be used with non-Newtonian fluids (10). A multiple-spindle technique is used. The multiple spindles permit the velocity profile to be reproduced. The nonlinear profile is determined by equipping the rotational viscometer with a series of spindles, each with a radius only slightly larger (an increase in radius of 0.125 inch) than the next smaller one. A series of tests are run on the fluid, each with a different spindle. The rotational speed (velocity) is varied until the same torque is reached in each test. Thus a series of points in the non-Newtonian velocity gradient are determined.

*Float-Type Viscometers.* The rotameter was discussed in the section on flow measurement. It was pointed out that flow could be measured by a float in a vertical tapered tube, but that the measurement was sensitive to changes in viscosity. In the viscometer, the weight of the float is balanced by the drag of the fluid through the tapered tube. If the flow is precisely controlled, float displacement indicates viscosity (2, 9).

b. Repeatability of 1% of reading
c. Accuracy of 2% below 35 centistokes; 4% above 35 centistokes
G. Vibrating-reed and ultrasonic viscometers (see Fig. 5-15)
  1. Theory of operation
    a. Probe is inserted in fluid and vibrated at specified frequency
    b. Viscous drag of fluid damps amplitude of vibration
    c. Damping is measured and related to fluid viscosity

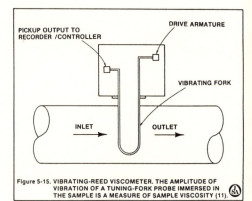

Figure 5-15. VIBRATING-REED VISCOMETER. THE AMPLITUDE OF VIBRATION OF A TUNING-FORK PROBE IMMERSED IN THE SAMPLE IS A MEASURE OF SAMPLE VISCOSITY (11).

H. Consistency
  1. Terminology defined — resistance to deformation of liquids (flowability)
  2. Factor affecting consistency measurements — temperature
  3. Methods of consistency measurements
    a. Laboratory measuring devices
      (1) Consistency cups (see Fig. 5-16)
        (a) Construction
          (1) Four brass cups — fixed orifices
          (2) Supporting rack

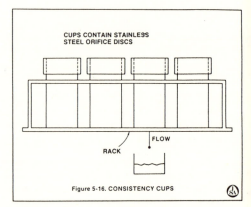

Figure 5-16. CONSISTENCY CUPS

        (b) Theory of measurement — time required for fixed volume of fluid through a fixed orifice
        (c) Applications of measurement
          (1) Paints — lacquers, enamels
          (2) Drilling fluids
          (3) Gelatin solutions
          (4) Petroleum industry
      (2) Calibrated pencil (see Fig. 5-17)
        (a) Constructed — calibrated aluminum pencil
        (b) Theory of measurement — with tip of pencil at fixed distance it is dropped for average number of readings of penetration

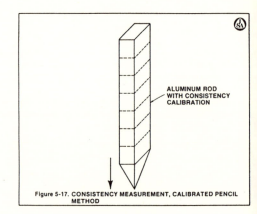

Figure 5-17. CONSISTENCY MEASUREMENT, CALIBRATED PENCIL METHOD

        (c) Applications of measurement — paper and pulp industries
      (3) Graduated troughs (see Fig. 5-18)
        (a) Construction
          (1) Trough (graduated)
          (2) Gate
        (b) Theory of measurement
          (1) Sample in trough behind gate
          (2) Distance fluid flows (under own weight) in a given time is measured by graduated scale

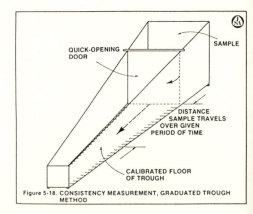

Figure 5-18. CONSISTENCY MEASUREMENT, GRADUATED TROUGH METHOD

*Vibrating-Reed Viscometers.* This unit continuously measures the viscosity of a liquid by measuring the damping effect or viscous drag of the liquid on a vibrating-reed probe (2, 9, 11).

The sensor is a probe, or tuning fork, of stainless steel. The probe is exposed to the process liquid and is connected to a reed section of magnetostrictive iron. This metal changes physical dimension with the strength of the magnetic field that surrounds it. Concentric with the reed assembly is a probe coil.

A current pulse to the coil changes the magnetic field around the reed, causing it to change dimension suddenly. The probe is shocked into vibration. The viscous material causes the vibration to die out with a damped frequency, where the damping is proportional to the viscosity. Electronic circuitry delivers another current pulse to the reed when the vibration drops to zero. The average current resulting from the driving pulses is measured and calibrated in terms of viscosity (2).

An alternative design of the vibrating-reed viscometer mechanically vibrates the probe in the fluid like a tuning fork. Vibration is continuous, and frequency is constant. The viscous liquid then damps the vibration by reducing its amplitude. Amplitude is then measured and calibrated in terms of viscosity (11).

*Ultrasonic viscometers* are similar to vibrating reed viscometers, wherein vibrations are induced in a metal blade used as a probe into the process fluid (9).

The probe or blade of the ultrasonic viscometer consists of magnetostrictive material. An ultrasonic electrical pulse excites longitudinal vibrations in the probe. The vibrations are dampened by the viscous fluid. The damping is electronically measured and calibrated in terms of viscosity (12).

The ultrasonic viscometer has been used successfully in determining the concentration of a high molecular weight polymer dissolved in a monomer or solvent. It is about one-tenth as sensitive to polymer molecular weight changes as is the rotational viscometer. It is very responsive in suspension and phase-changing polymers (12).

(c)  Applications of measurement
    (1)  Jams, preserves
    (2)  Tomato products
(4)  Rotational viscometers (for consistency measurement)
    (a)  Construction
       (1)  Spindle assembly — spring-loaded
       (2)  Rotating table assembly
       (3)  Indicating instrument
    (b)  Theory of measurement — resistance to deformation by a fluid is directly proportional to the torque on spindle
    (c)  Applications of measurement
       (1)  Paper coating
       (2)  Catsup manufacturing
       (3)  Starch sizing
       (4)  Ink — rotogravure presses

b.  Industrial measuring devices
(1)  Drainage rate type
    (a)  Theory of measurement — based on rate of drainage of solvent from solids in suspension
    (b)  Method of measurement
       (1)  Consistency detector collects fibers or fluid
       (2)  Detector box head pressure activate pressure signal — controls water flow rate
    (c)  Application — paper and pulp industries

(2)  Electrical resonance type (see Fig. 5-19)
    (a)  Theory of measurement — fluids will vary in frequency dependent upon amount of solid material in mixture
    (b)  Method of measurement
       (1)  Body of pure water in high-frequency circuit (250 megacycles)
       (2)  Addition of fibrous material will change resonant frequency
       (3)  Amount of frequency shift will determine consistency measurement
    (c)  Application — paper and pulp industry

(3)  Apparent viscosity type
    (a)  Mechanical
       (1)  Theory of measurement — torque required to drive feeler device (paddle, spindle, etc.) will vary

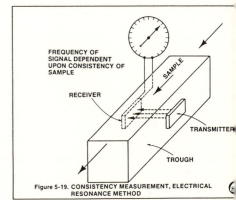

FREQUENCY OF SIGNAL DEPENDENT UPON CONSISTENCY OF SAMPLE

SAMPLE

RECEIVER

TRANSMITTER

TROUGH

Figure 5-19. CONSISTENCY MEASUREMENT, ELECTRICAL RESONANCE METHOD

dependent upon consistency of fluid

    (2) Open-box type — open container measurements

    (3) Closed-pressure type — pipeline measurements

(b) Hydraulic

    (1) Theory of measurement — variation of viscosity while under constant flow produces variations in head of fluid; variation indicates degree of consistency

    (2) Float-level type

    (3) Dual type

    (4) Slope type

(c) Applications of apparent viscosity measurements

    (1) Paper stock control

    (2) Mining

    (3) Food processing

(4) Optical type (see Fig. 5-20)

(a) Theory of measurement — the amount of radiant energy that will pass through a fluid determines its consistency

(b) Components

    (1) Continuous sampling chamber

    (2) Light source

    (3) Photocell

    (4) Amplifier and indicator

(c) Application — pulp consistency measurement

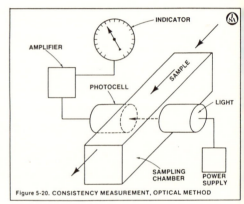

Figure 5-20. CONSISTENCY MEASUREMENT, OPTICAL METHOD

III. Thermal conductivity

    A. Absolute steady-state measurement (see Fig. 5-21)

        1. Guarded hot plate

           a. Theory of operation

              (1) Two samples are placed adjacent to a flat plate heater assembly

              (2) The heater is surrounded by a guard heater that prevents radial heat flow

              (3) Heat flow from main heater to liquid heat sink travels through the samples

              (4) Thermocouples on each side of sample measure temperature gradient

              (5) Temperature gradient is measurement of thermal conductivity

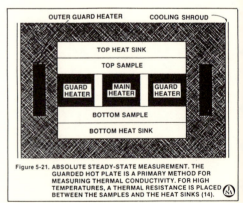

Figure 5-21. ABSOLUTE STEADY-STATE MEASUREMENT. THE GUARDED HOT PLATE IS A PRIMARY METHOD FOR MEASURING THERMAL CONDUCTIVITY. FOR HIGH TEMPERATURES, A THERMAL RESISTANCE IS PLACED BETWEEN THE SAMPLES AND THE HEAT SINKS (14).

Thermal Conductivity: This measurement can be defined quite simply, but in practice accurate measurements are difficult to carry out. Both steady-state and dynamic thermal conductivity measurements are commonly made. Steady-state measurements are based on the one-dimensional equation for conductive heat transfer. Primary steady-state measurements of conductivity are made by power dissipation in an electrical heater, by temperature measurement in a constant flow of water, or by the boiling off of a liquid of known thermal properties (14).

One primary absolute method of steady-state measurement is that of the guarded hot plate. Two samples are placed on either side of a heater (hot plate) assembly. The heater assembly is made up of an inner heater surrounded by an annular outer (guard) heater. The guard heater eliminates radial heat flow from the inner heater. Thus one-dimensional heat flow through the sample is produced. Liquid-cooled heat sinks on the opposite sides of the sample provide a temperature gradient across the sample.

Thermocouples are installed on each surface of the sample. The temperature difference across the sample is then indicative of the thermal conductivity (14).

The thermal conductivity gas analyzer provides a dynamic measurement. It consists of two wires, usually platinum, each of which is a leg of a Wheatstone bridge circuit. Each wire is stretched through a measuring cell. A reference gas is placed in one measuring cell, while the gas sample flows through the second cell. Thus one filament (wire) is exposed to the reference gas and the other to the sample gas. This arrangement provides a comparative measurement between reference and sample gas (15).

Since thermal conductivity is the ability of a fluid transfer heat, a gas that transfers heat from a source at a greater rate than another gas has a higher thermal conductivity. The dual conductivity cell therefore compares the thermal conductivity of an unknown gas to that of a known reference gas (2). The wire filament in each side of the cell is heated to a low temperature by an electrical current. Heat is lost from the filament to the gas in each side of the cell. The heat loss varies as the conductivity of the gas. Heat loss changes filament temperature and filament resistance varies with temperature. The resistance is then measured by the bridge circuit by comparison to the resistance of the filament in the reference gas of known conductivity.

The conductivity analysis is usually made to identify the volumetric percentage of one gas in a gaseous mixture. Identification and calibration are made possible by the choice of the reference gas. Actually, nonsteady-state methods determine diffusivity, which is defined as the ratio of thermal conductivity to specific heat. Since diffusivity is being measured, the dynamics of the conductivity measurement vary with the gas composition (16). Thus a response time of 5 seconds is typical for a hydrogen measurement, while 25 seconds would be typical for carbon dioxide.

The resistance measurement tends to drift with time because of the vaporization of the platinum filament, and in some cases because of the reaction of the hot filament with the gas. The filament is commonly coated with glass to eliminate the drift. The coating, however, tends to further slow the response of the measurement.

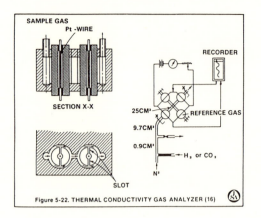

Figure 5-22. THERMAL CONDUCTIVITY GAS ANALYZER (16)

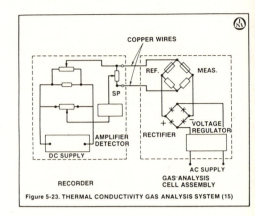

Figure 5-23. THERMAL CONDUCTIVITY GAS ANALYSIS SYSTEM (15)

b. Experimental precautions
   (1)  Ratio of main heater size to sample thickness must be great enough to minimize heat loss from periphery of sample
   (2)  Thermocouples must not be so large as to affect heat flow
   (3)  Temperature difference between guard and main heaters must be less than 0.3°C
   (4)  Accurate heater power measurements are required
   (5)  Accuracy +20% to –10%
B. Dynamic analyzers for gases (see Figs. 5-22 and 5-23)
  1. Theory of operation
    a. Thin, heated platinum wires are stretched through two gas chambers
    b. One chamber contains a reference gas of known conductivity, and gas sample flows through the other
    c. Platinum wires are heated; heat loss to each gas is proportional to thermal conductivity of the gas
    d. Temperature of wire varies with heat loss; resistance varies with temperature
    e. Wire resistance is measured on a bridge-type instrument, which is calibrated in terms of thermal conductivity
  2. Operating characteristics
    a. Wire filament responds in 0.1 sec
    b. Rate of heat loss by wire varies with diffusion through gas
    c. Each gas had different diffusion rate; thus cell response varies from gas to gas

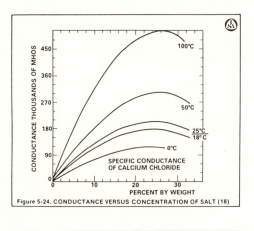

Figure 5-24. CONDUCTANCE VERSUS CONCENTRATION OF SALT (18)

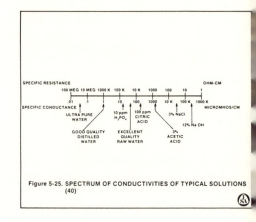

Figure 5-25. SPECTRUM OF CONDUCTIVITIES OF TYPICAL SOLUTIONS (40)

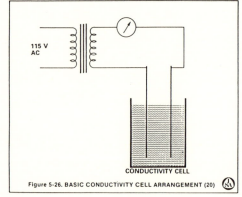

Figure 5-26. BASIC CONDUCTIVITY CELL ARRANGEMENT (20)

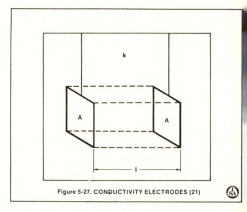

Figure 5-27. CONDUCTIVITY ELECTRODES (21)

## IV. Electrical conductivity

A. Theory of conductivity measurement (see Fig. 5-28)
1. Conducting ability of ions present in a solution
2. Basically a measure of electrical resistance of a solution — circuit diagram (Wheatstone bridge)
3. Polarization of electrodes — ac rather than dc current employed in conductivity measurements due to reduced errors (Kohlrausch)

B. Units of measurement
1. Micromho (mho) conductance of a solution by which a potential difference of one millionth of a volt will cause a current flow of one millionth of an ampere
2. Specific conductance — standard unit of measure — conductance in mhos of one cubic centimeter of solution measured between two electrodes one centimeter apart (calculated at 77°F = 25°C)
3. The cell constant — relates actual measured value to the standard value (ratio of specific conductance to measured conductance)

C. Conductivity cells (see Fig. 5-29)
1. Types
   a. Flow
   b. Immersion

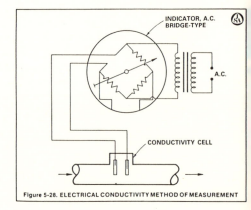

Figure 5-28. ELECTRICAL CONDUCTIVITY METHOD OF MEASUREMENT

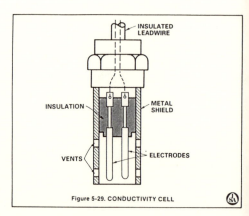

Figure 5-29. CONDUCTIVITY CELL

Electrical Conductivity: The earliest attempt to use solution conductivity as an industrial measurement was made in 1910 in determining the purity of distilled water, The first sensors were similar to fragile laboratory equipment, but developments over the years have resulted in sturdy and stable instruments, able to withstand the demanding environments of process streams. The modern use of stainless steel and high-density alumina insulating materials has resulted in a rugged assembly that can be used in pressures of 10,000 psi. Removal devices such as packing glands and gate valves permit the cell to be removed from, and reinserted into, a process stream under line pressure and temperature conditions (18).

Conductivity measurement is the determination of a solution's ability to conduct an electric current. Pure water is a nonconductor and has a conductivity of essentially zero. When an electrolyte (a material that ionizes such as an acid or a base) is dissolved in the water, the solution will have a unique conductivity, depending upon the electrolyte, the concentration, and the temperature of the solution. A measurement of the conductivity is thus a measurement of the concentration of the electrolyte in the solution at the given temperature (19). Fig. 5-24 shows the variation of the conductance of a salt versus the concentration of the solution.

The strong effect of temperature on conductivity is illustrated by Fig. 5-24. The sensitivity of conductivity to temperature is called the temperature coefficient of conductivity. The coefficient can vary greatly, depending on the nature and concentration of both the electrolyte and the non-ionic materials in the solution. Increasing temperatures almost always increases conductivity, and thus most conductivity measurements must be temperature compensated (18). Temperature compensation means that the compensated reading is that which would be obtained if the solution were at some reference temperature, usually chosen as 25°C.

Conductivity is measured by immersing two conducting electrodes in a liquid and applying an ac voltage across the electrodes. Alternating current is required since direct current will cause polarization of the electrodes and subsequent high resistance at the electrode surfaces. The high resistance will then mask the conductivity measurement. The voltage forces a current to flow between the electrodes, as shown in the circuit illustrated in Fig. 5-26.

The resistance provided by the ionized solution to the electrical current between the electrodes is measured by a Wheatstone bridge circuit, where it is converted to units of conductivity. A definition of specific conductivity (18) is the "reciprocal of the resistance in ohms measured between opposite faces of a centimeter cube of an aqueous solution at a specific temperature." The "cube" spoken of is made up of the area of the electrodes multiplied by the distance between them, as seen in Fig. 5-27.

Conductivity is proportional to the cross-sectional area (A) of the cube, and inversely proportional to the distance across it (l). Obviously, trade-offs can be made betweeen area and distance to provide the same value of conductivity.

2. Materials of cells
   a. Hard rubber — alkaline solutions
   b. Glass — most chemicals with exception of alkalis and fluorides
3. Electrodes (see Fig. 5-30)
4. Cell connecting cable — copper, double conductor, rubber coated
   a. Gage — dependent on recorder range span
   b. Length — no effects up to 500 feet
5. Cell requirements
   a. Electrical
      (1) Limits of measurement — 50-100,000 ohms
      (2) Cell constants — 0.01 to 100
   b. Chemical — resistant to chemical action
   c. Mechanical
      (1) Temperature — glass cells, to 300°F
      (2) Pressure — glass cells, 50 psi
6. Calibration check of cell — check coil for recorder
7. Cell maintenance — electrode care
   a. Intermittent use — storage in water, rinsing with distilled water
   b. Sensitizing electrodes
      (1) Hydrochloric, nitric acid — 10% solution
      (2) Replatinizing

D. Electrodeless conductivity cells (see Figs. 5-31 and 5-32)
1. Theory of operation
   a. Two toroidally wound coils encircle an electrically nonconductive tube
   b. A loop of solution couples the two toroids
   c. An ac signal is applied to one (primary) toroid
   d. A current is induced in the loop of solution
   e. Current varies with conductance of the solution
   f. Current induces a current in second toroid, which is proportional to conductivity
2. Characteristics
   a. Range 50 to 1000 millimhos
   b. Used with solutions that are abrasive, corrosive, or contain entangling fibers
   c. Applications
      (1) Ore leaching
      (2) Seawater
      (3) Raw sewage
      (4) Bauxite
      (5) Radioactive solutions
      (6) Sulfuric acid

E. Factors affecting measurement
1. Temperature of the solution
   a. Manual compensation — variable resistance as one leg of bridge circuit

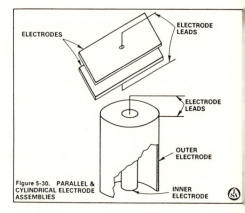

Figure 5-30. PARALLEL & CYLINDRICAL ELECTRODE ASSEMBLIES

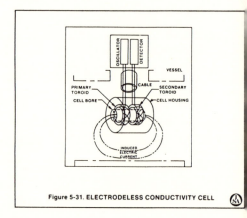

Figure 5-31. ELECTRODELESS CONDUCTIVITY CELL

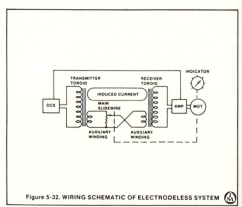

Figure 5-32. WIRING SCHEMATIC OF ELECTRODELESS SYSTEM

Conductivity measurement system performance requires that the electrode remain clean. Measurement errors are introduced by the fouling of the electrodes due to coating by oil, algae, and scale, or by entanglement by fibrous materials (40). The electrodeless conductivity sensor is used where fouling is probable. In this instrument, the two electrodes are replaced by two electrically coupled toroids.

The toroidally wound coils encircle an electrically nonconductive tube that carries the process liquid. The two coils can be viewed as two transformers: The first transformer is made up of the input coil as a primary winding with the solution loop as the secondary winding; and the second transformer has the solution loop as the primary winding with the second coil as the secondary or output winding. Thus a loop of solution couples the two toroids.

An ac voltage in the audio frequency range is supplied to the input coil, and a detection device is connected to the output coil. A constant frequency and voltage applied to the input coil will cause a current to be generated in the loop of solution.

The current varies directly as the conductance of the solution. This current induces a current in the second (output) coil, which is in turn proportional to the conductance of the process liquid (40, 41, 42).

Electrodeless conductivity is applicable to the higher conductivity ranges such as 50 to 1000 millimhos per centimeter. A transformer bridge is used in practice to reduce errors introduced by fluctuations in excitation signal and voltage. The four legs of the bridge are made up of the two toroid coils described above plus two more coils introduced to null the bridge.

Conductivity, like viscosity, turbidity, and thermal conductivity is an inferential measurement; that is, in itself, it carries no information about the ionic carriers of the current. Therefore, a measurement of concentration requires that the conductivity-concentration curve (Fig. 5-24) must be known in advance. For this reason, most practical industrial applications of conductivity have been in the monitoring of water purity. Examples have been quality measurements for boiler feedwater and detection of contamination due to acid leaks in process coolers. Efforts are being made to extend the usefulness of the measurement to determining the extent of reactions involving ionizable materials and the mixing effects, in the same type solutions (22).

Experimental work has indicated the same care must be exercised in the monitoring of electrolytes dissolved in water (23, 43). The likelihood of obtaining misleading information is especially great when dealing with relatively low concentrations of unbuffered solutions. The problem is in the extreme mobility of the hydrogen ion, which causes relatively small changes in pH to effect conductivity greatly, so that the calibration curves relating conductivity to concentration become unreliable.

   b. Automatic compensation
      (1) Compensating cell — fixed concentration
      (2) Temperature compensating coil — as one
          leg of bridge circuit
   2. Effective area of electrodes — need for cell
      maintenance
   3. Potential difference and distance between
      electrodes
 F. Applications of electrolytic conductivity measuring
    devices
    1. Plating and pickling operations
    2. High-pressure steam plants — steam purity, boiler
       blowdown, feedwater
    3. Acid and base concentrations — nitrocellulose
       manufacture
    4. Chemical processing
    5. Dye industries

V. Oxygen analysis
   A. Paramagnetic analyzers
      1. Paramagnetic properties of gases — those which
         characteristically are attracted into a magnetic field
      2. Paramagnetic gases
         a. Oxygen
         b. Nitrogen oxides
      3. Types of oxygen analyzers
         a. Paramagnetic (see Figs. 5-34 and 5-36)
            (1) Principle of operation — indicators and
                recorders
                (a) Small glass dumbbell-like sphere is
                    suspended in a non-uniform magnetic
                    field by a quartz fiber
                (b) Spheres will remain in a balanced
                    position when oxygen is not present
                    (quartz fiber torque or torsional force
                    is balanced by the magnetic force of
                    the pole pieces)
                (c) Upon the introduction of a sample
                    (containing traces of oxygen) the
                    magnetic field is changed
                (d) The magnetic force, proportional to
                    the oxygen content, will cause sphere
                    (test body) to rotate — degree of
                    rotation proportional to oxygen
                    content
         b. Paramagnetic — thermal conductivity
            combinations (see Fig. 5-35)
            (1) Principle of operation
                (a) Electrically heated matched filaments

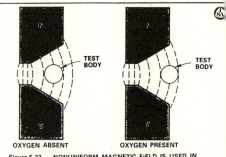
Figure 5-33. NONUNIFORM MAGNETIC FIELD IS USED IN PARAMAGNETIC OXYGEN ANALYZERS. WHEN OXYGEN IS NOT PRESENT, TEST BODY ASSUMES A CERTAIN POSITION. WHEN OXYGEN IS PRESENT, IT IS ATTRACTED INTO THE MAGNETIC FIELD, DISPLACING THE TEST BODY. AMOUNT OF OXYGEN CAN BE DETERMINED BY AMOUNT OF DISPLACEMENT. (20

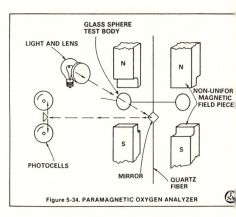
Figure 5-34. PARAMAGNETIC OXYGEN ANALYZER

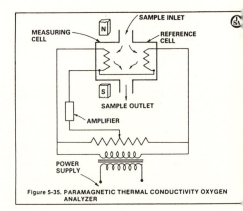
Figure 5-35. PARAMAGNETIC THERMAL CONDUCTIVITY OXYGEN ANALYZER

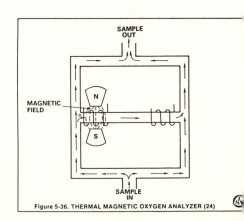
Figure 5-36. THERMAL MAGNETIC OXYGEN ANALYZER (24)

Oxygen Analysis: The principles by which oxygen analyzers are designed and applied are as follows: (1) magnetic susceptibility, (2) catalytic, (3) electrolytic, and (4) solid state.

Magnetic susceptibility is a measure of the intensity of the magnetization of a substance when placed in a magnetic field. All materials are susceptible, and iron is an example of extreme susceptibility. Some materials tend to become slightly magnetized along with the magnetic field and are said to be *paramagnetic*. Oxygen is a paramagnetic material, which means that it is drawn into a magnetic field. The magnetic susceptibility of oxygen is several hundred times greater than that of most other gases (24).

An analyzer designed to measure the magnetic susceptibility of a gas mixture is essentially an oxygen analyzer. Several analyzers are manufactured using this principle. They vary in component arrangement and operating characteristics. One is called a *thermal magnetic oxygen analyzer*. It divides the gas sample into two equal streams in the measuring cell. A passage is provided connecting the two sample streams. With the divided sample flow being identical, there is normally no sample flow through the passage. An electrical heating filament is placed at each end of the crossover passage. The filament at one end has a strong magnet around the heating filament. Any oxygen in the flowing sample will be attracted to the magnetic field. The paramagnetic susceptibility varies inversely as the square of the temperature and thus decreases rapidly as temperature is increased. The sample is heated by the electrical filament, losing its susceptibility. It is then displaced by cool oxygen being attracted to the magnetic field. This action cools the filament in the magnetic field, causing its resistance to be different from that of the other electrical heating filament. The difference in the resistance is measured on a bridge-type instrument and is calibrated in terms of percent oxygen.

The second oxygen analyzer based on the magnetic susceptibility of oxygen utilizes a miniature, sensitive torsion balance made up of a small dumbbell-shaped body suspended on a metallic or quartz fiber. It is similar to the instrument used to measure gas density as described in the section on density and is called a torsion balance analyzer. The dumbell body is free to rotate in a plane in the space between the poles of a magnet. The dumbbell body is magnetically susceptible, and each ball is attracted to the point of maximum field strength. If oxygen is present in the gas sample introduced into the cell containing the dumbbell, it will also be attracted to the point of maximum field strength. It will in this manner displace the ball of the dumbbell as shown in Fig. 5-33. The oxygen displaces the balls of the dumbbell, causing it to rotate. The rotation is in proportion to the amount of oxygen in the sample. Rotation is measured by a beam of light on a mirror attached to the dumbbell and is calibrated in ppm of oxygen (24).

separated by nonmagnetic metal
- (b) Equal temperatures — like resistance values
- (c) Magnetic flux positioned around one filament — oxygen attraction to flux density
- (d) Increased temperature of sample gas — loses magnetic qualities
- (e) Sample gas exchange — cools filament (resistance change)

4. Factors affecting measurement
   a. Temperature variations — thermostat
   b. Pressure variations — pressure resistance transducer
   c. Flow rate of sample

5. Application of magnetic analyzers
   a. Combustion control by determining proper air supply for maximum fuel efficiency
   b. Protective atmospheres to reduce oxidation — metal and food industries
   c. Process control for reactivating catalyst beds
   d. Monitor combustible atmospheres
   e. Medical — oxygen measurement and metabolism studies
   f. Aerospace — missile fuel loading
   g. Process gas stream monitoring and control
   h. Air pollution analysis

B. Catalytic analyzers
   1. Theory of operation (see Fig. 5-37)

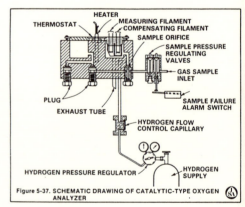

Figure 5-37. SCHEMATIC DRAWING OF CATALYTIC-TYPE OXYGEN ANALYZER

   a. A constant pressure sample is brought into the analyzer block
   b. Gas sample is mixed with hydrogen at a constant temperature
   c. Major portion of gas sample flows into measuring chamber where it burns, raising the temperature and the electrical resistance of the measuring filament
   d. Small portion of gas sample flows into compensating chamber and surrounds the compensating filament; thus physical properties of gas have the same effect on both filaments
   e. Filaments are legs of a Wheatstone bridge, which is sensitive to measuring filament temperature, but not to gas physical properties, which are cancelled out by compensating filament

The catalytic oxygen analyzer mixes the gas sample with hydrogen, then passes the mixture over a coated, heated filament that acts as a catalyst to produce combustion. The filament is a noble metal catalyst filament that causes combustion of the gas sample by its initial temperature. Combustion raises the filament temperature, causing a change in its resistance. The resistance change is measured by a bridge-type instrument calibrated in terms of percent oxygen (2).

2. Applications
   a. Flue gas analysis, to assure adequate oxygen for combustion
3. Operation
   a. A continuous hydrogen supply must be assured

C. Electrolytic analyzers (see Fig. 5-38)
1. Theory of operation
   a. Sensor is galvanic cell consisting of two electrodes, one of which is exposed to the sample gas
   b. Electrochemical reaction occurs at interface of silver cathode and electrolyte; and oxygen is converted into hydroxyl ions
   c. The reaction results in a flow of electrical current through the cadmium anode
   d. The current is proportional to the oxygen content of the gas sample
2. Applications
   a. Measurement of traces of oxygen in a gas
3. Operation
   a. Range from 0 to 5 ppm to 0 to 10,000 ppm
   b. Nonlinear output for wide ranges
   c. Less sensitive at high concentrations
   d. Accuracy ±2%

D. Solid-state analyzer (see Fig. 5-39)
1. Theory of operation
   a. Oxygen in process gas comes in contact with ceramic oxide which, when hot, becomes an electrolytic conductor exclusively for oxygen ions
   b. Electrical output is produced; proportional to the difference in the amount of oxygen on either side of cell
   c. A reference gas, usually air, is introduced to one side of the cell, usually the inside of the cylindrical probe
   d. Cell output is then related to the amount of oxygen in the gas sample
2. Applications
   a. Direct insertion in flowing gas
   b. Monitors combustion mixtures, combustion products, and process streams
3. Operation
   a. Range 0-0.1 to 0-10% oxygen
   b. Accuracy 0.1% of reading
   c. Operating temperature range 10-760°C

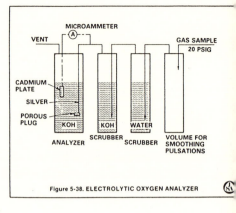

Figure 5-38. ELECTROLYTIC OXYGEN ANALYZER

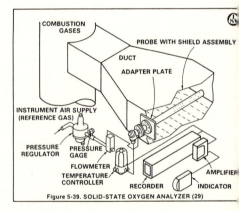

Figure 5-39. SOLID-STATE OXYGEN ANALYZER (29)

The electrolytic oxygen analyzer requires scrubbing of the gas sample to remove impurities, further scrubbing by potassium hydroxide to remove acid gases, and humidifying the sample. The sample then flows to a galvanic cell consisting of two electrodes immersed in an electrolyte. An electrochemical reaction takes place at the interface of the cathode and the electrolyte, whereby the oxygen is converted into hydroxyl ions. The ions then flow to the cadmium plate. The combination of the hydroxyl ions with the cadmium produces electrons, which flow as a current through a microammeter calibrated in units of parts/million of oxygen. The electrolytic analyzer is a very sensitive device. but workability is highly dependent on equipment design (27).

The solid-state oxygen analyzer probe (25) is applied directly to the gas stream and requires no sampling system, as do the oxygen analyzers formerly discussed. The heart of the probe is a stabilized zirconium cell that is heated to a controlled temperature of 850°C. At this temperature, it generates a voltage that is related to the difference in the partial pressures of oxygen in the process gas stream and in a reference gas. The reference gas is applied to one side of the cell and the process gas to the other. The lower the oxygen concentration in the process gas, the greater the voltage generated.

The insertion of the probe directly into the process gas stream results in a more representative oxygen sample being applied to the probe and eliminates the continuous maintenance on the sampling system.

E.  Dissolved-oxygen sensors
  1. Electrolytic sensor (see Fig. 5-40)
    a.  Theory of operation
      (1) Oxygen passes through membrane at a rate proportional to the amount of oxygen in the water outside the membrane
      (2) Electrochemical reaction of oxygen with cathode produces hydroxyl ions in the electrolyte
      (3) Ions migrate to the anode, causing a current to flow
      (4) Current flows through load resistor in electrical circuit; voltage across resistor is related to oxygen in water
    b.  Applications
      (1) Measurement of oxygen in natural water sources
    c.  Operation
      (1) Ranges 0-20 ppb to 0-25 ppm
      (2) Accuracy ±5%, for scale 0-20 ppb; ±1%, for scale 0-25 ppm
      (3) Sample temperature 0-50°C
  2. Boiler feedwater sensors
    a.  Electrode sensor (see Fig. 5-41)
      (1) Theory of operation
        (a) Water sample flows through cell
        (b) Oxygen in sample reacts with electrodes
        (c) Electrical current flows which is proportional to the amount of oxygen in the sample
      (2) Operation
        (a) Range 0-25 ppb to 0-250 ppb
        (b) Accuracy ±5%
    b.  Chemical reaction sensor (see Fig. 5-42)
      (1) Theory of operation
        (a) Oxygen in sample reacts with nitric acid in reaction column
        (b) Products of reaction increase electrical conductance of sample
        (c) Change in conductance is measured and related directly to concentration of dissolved oxygen
      (2) Operation
        (a) Range down to 0-1 ppb
VI. pH (hydrogen ion concentration)
  A.  Characteristics
    1. Definition of pH — effective acidity or alkalinity of a solution

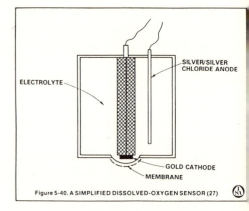

Figure 5-40. A SIMPLIFIED DISSOLVED-OXYGEN SENSOR (27)

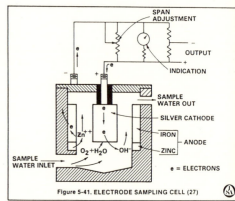

Figure 5-41. ELECTRODE SAMPLING CELL (27)

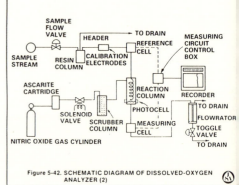

Figure 5-42. SCHEMATIC DIAGRAM OF DISSOLVED-OXYGEN ANALYZER (2)

*Dissolved oxygen.* A widely accepted measurement of water quality is the dissolved oxygen content of natural streams. A small amount of oxygen will dissolve in water through natural processes such as photosynthesis. This oxygen purifies the water by the bio-oxidation of the organics and other oxygen-demanding wastes. However, since the amount can be dissolved is small, it can be rapidly depleted by an overburden of pollutants, leaving the water with an unsightly appearance, offensive odor, and expired aquatic life (26).

Most sensors used for the measurement of dissolved oxygen in natural processes work on the principle of the electrolytic cell, formerly described. Two electrodes, a gold or platinum cathode and a silver anode, are contained in a chamber filled with electrolyte. The chamber is a plastic cell with a semipermeable membrane to separate the electrolyte from the medium to be measured. This construction isolates the electrodes and the electrolyte from the water in which the sensor is immersed. The oxygen migrates through the membrane to reach the cathode. An electrochemical reaction takes place at the cathode that turns the oxygen in the cell into hydroxyl ions, which travel to the anode. The ions cause a current flow in the electrical circuit connecting the electrodes, and the current is proportional to the oxygen reacted.

Since the oxygen reacts in the cell, the oxygen pressure inside the membrane is essentially zero. The oxygen pressure in the water on the outside of the membrane causes it to diffuse through the membrane into the cell. Thus the amount of oxygen reacted in the cell depends on the pressure exerted on the membrane by the oxygen in the water. In this manner, the electrical current is proportional to the dissolved oxygen (26, 27).

While the measurement of dissolved oxygen in natural waters is taken to assure its presence, the measurement of dissolved oxygen in boiler feedwater is made to assure that the amount is a minimum. Minute traces of oxygen in the feedwater cause corrosion in steam generating equipment and must be detected and removed.

The dissolved oxygen sensor for feedwater service is based on the chemical reaction of oxygen with nitric oxide. The products of the reaction increase the electrical conductivity of the sample. The measurement of the change in conductance of the sample is then calibrated in terms of the amount of dissolved oxygen.

pH Sensors: The concept of pH was introduced on 1909 by S.P.L. Sorenson (28). He defined pH as:

$$pH = Log_{10} \, C_H \qquad\qquad\qquad (5\text{-}1)$$

where $C$ = concentration.

Quite simply, pH is the measurement of the acidity or alkalinity of a liquid that contains a proportion of water, just as the "inch" is a measurement of length and "degree" a measurement of temperature. Equation 5-1 shows the pH to be related to the concentration or the strength of the liquid solution. Thus it permits the assignment of a definite, exact value of pH to a specific concentration. The advantages of the definite relationship are well illustrated by the commercial manufacture of jelly in the food industry (30). For many years, the making of jelly was a haphazard "art," and some batches would come out thick, some thin, and some would fail to jell at all. Then the following relationship between final product characteristics and pH was found:

|       | pH  | Condition of jelly      |
|-------|-----|-------------------------|
| Below | 2.6 | Refuses to jell         |
|       | 2.6 | White precipitate forms |
|       | 2.8 | "Weeping" occurs        |
|       | 3.1 | Maximum stiffness       |
|       | 3.2 | Medium stiffness        |
|       | 3.3 | Tender                  |
|       | 3.4 | Very tender             |
| Above | 3.5 | Refuses to jell         |

The measurement of pH thus makes possible the manufacture of uniformly consistent jelly with desired characteristics.

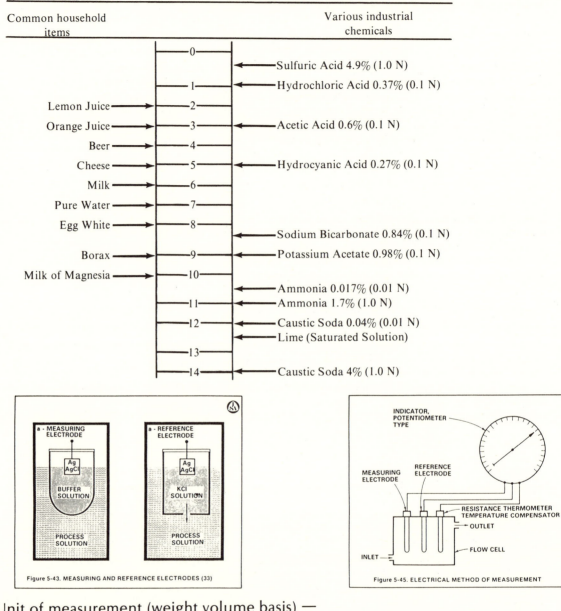

Table 5-3
pH Values of Various Materials (27)

| Common household items | | Various industrial chemicals |
|---|---|---|
| | 0 | Sulfuric Acid 4.9% (1.0 N) |
| | 1 | Hydrochloric Acid 0.37% (0.1 N) |
| Lemon Juice | 2 | |
| Orange Juice | 3 | Acetic Acid 0.6% (0.1 N) |
| Beer | 4 | |
| Cheese | 5 | Hydrocyanic Acid 0.27% (0.1 N) |
| Milk | 6 | |
| Pure Water | 7 | |
| Egg White | 8 | Sodium Bicarbonate 0.84% (0.1 N) |
| Borax | 9 | Potassium Acetate 0.98% (0.1 N) |
| Milk of Magnesia | 10 | Ammonia 0.017% (0.01 N) |
| | 11 | Ammonia 1.7% (1.0 N) |
| | 12 | Caustic Soda 0.04% (0.01 N) |
| | | Lime (Saturated Solution) |
| | 13 | |
| | 14 | Caustic Soda 4% (1.0 N) |

Figure 5-43. MEASURING AND REFERENCE ELECTRODES (33)

Figure 5-45. ELECTRICAL METHOD OF MEASUREMENT

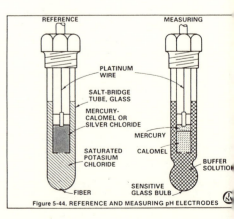

Figure 5-44. REFERENCE AND MEASURING pH ELECTRODES

a. Unit of measurement (weight volume basis) — the negative log of H ion concentration
b. pH values of acid and base solutions and extent of pH scale (0-14 pH)—relate to pure water (7 pH)
c. Applicable only to hydrous solutions
2. Methods of pH measurement
   a. Color indicating method (colormetric)
      (1) Limitations — time, accuracy
   b. Electrical method (potentiometric) (see Figs. 5-44 and 5-45)
      (1) pH cell construction and operation (relate to principle of wet storage battery) — an emf is developed that is proportional to the hydrogen-ion concentration of the solution

About the time that Sorenson was defining pH, G.N. Lewis introduced the concept that ionic activity was distinct from component concentration, and thus that the pH galvanic cell was measuring activity rather than concentration. The definition therefore became

$$pH = - \text{Log } H^+ \tag{5-2}$$

where $H^+$ is hydrogen ion concentration and the logarithm is to base 10.

The ion is simply a charged particle that determines the current-carrying capacity of the solution. The measurement of pH is thus the measurement of the dissociation of the acid or alkalai molecules into ions. Some acids dissociate readily and produce high concentrations of hydrogen ions. These are known as strong acids. Sulfuric acid is an example — it will readily dissolve iron nails. Boric acid, on the other hand, can be safely used as an eye wash because it is a weak acid and does not dissociate readily (30, 31).

Table 5-3 shows that pH measurement is meaningful only for dilute concentrations of an acid or alkalai dissolved in water. This limitation exists because only at high dilutions does the activity coefficient (the relationship between hydrogen ion concentration in Equation 5-2 and component concentration in Equation 5-1) approach unity. Thus the dissociation of water molecules controls all reactions in aqueous solutions. Water dissociates slightly (is a weak electrolyte) into equal numbers of hydrogen (acid) and hydroxyl (alkalai) ions and thus is defined as neutral (31, 32). Pure water has a pH of 7, as shown in Table 5-3.

Two practical methods of pH measurement are in use (31). One of these is the *colorimetric* method, which uses substances that change color when subjected to acidic or alkaline environments. One such substance is litmus paper, which is dipped into the liquid sample and changes to a different color depending on whether the sample is acidic or alkaline. A quantitative measurement is gained by the use of liquid indicators that change color at a specific pH. The concentration of hydrogen ion in the solution changes the constitution of the indicator and thus its color.

While measurements of pH to within 0.1 pH units are possible using colorimetric liquids in the laboratory, there are difficulties associated with the method:

(1) Addition of the indicator can alter the pH value of the sample.

(2) The measurement is masked by dissolved or suspended matter in the sample.

(3) Accuracy is affected by the color of the sample.

The second technique for the measurement of pH is the *electrometric* or *potentiometric* method. This is almost universally used in the processing industries. The pH is determined by measuring the voltage developed by two electrodes in contact with the electrolytic solution. As in a galvanic cell (a battery), a chemical reaction takes place between the electrodes and the ionized solution (electrolyte), producing an electrical potential. The measuring system consists of three major elements: the measuring electrode, the reference electrode, and an instrument to sense and amplify the very small voltage developed between the two electrodes (33). Figure 5-43 shows the measuring and reference electrodes.

The measuring electrode produces a dc voltage at its special glass surface that is proportional to the pH of the stream being measured. The glass surface is a specially formulated glass membrane that responds to the hydrogen and hydroxyl ions by an internal exchange of ions within the membrane. The membrane is fused to the glass body so that the outer surface makes contact with the process fluid while the inner surface is in contact with an internal filling solution. The internal electrical connection is made to the glass surface via a buffer solution, which in turn is in contact with a silver-silver chloride couple surrounding the lead wire. While measuring electrodes are made of a variety of other materials, this particular configuration has exhibited high thermal stability and long term stability (32, 33).

       (a) Measuring electrode ("pH-sensitive")
— glass — generates own potential
based on pH concentration of
solution; pH response takes place at
surface of glass membrane

       (b) Reference electrode — internal
components maintain constant
potential

          (1) Calomel

          (2) Silver/silver chloride

       (c) Salt bridge — potassium chloride

       (d) Cell output — algebraic sum of the
two electrode potentials

       (e) pH scale nonlinear with respect to
concentrations

       (f) Types — flow, immersion

    (2) Circuit characteristics

       (a) Voltage — 59.1 millivolts/pH at 25°C
(Nernst equation: voltage as a function
of pH)

       (b) Low current flow — approximately 1
micro-microampere

       (c) High resistance (50 to 500 megohms)

  3. Speed of response

    a. Time constant of from 1 to 2 seconds for clean
electrodes in a moving stream

  4. Accuracies: generally ±0.02 pH

  5. Factors affecting pH measurement

    a. Temperature — change in degree of ionization
with increase in temperatures (operating range:
0-100°C) compensation necessary for accuracy
in pH meassurement

      (1) Manual rheostat

      (2) Resistance thermometer

    b. Pressure — under pressure solution may enter
reference electrode and contaminate inner
cell; sealed electrode assemblies utilized for
solutions under pressure (normal pressure
range: 0-30 pounds/square inch)

    c. Velocity — more than 0.1 foot/second to
maximum of 10 ft/second

    d. Chemicals — oils, tars, abrasives, solids,
hydrofluoric acids

B. Configurations of pH-measuring cells

  1. Encapsulated with protected electrodes (see Fig.
5-46)

    a. Epoxy encapsulation

    b. Temperature limits — 5 to 65°C

  2. Self-cleaning (see Fig. 5-47)

    a. Flat surface flush with conduit

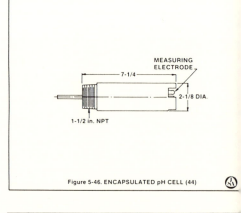

Figure 5-46. ENCAPSULATED pH CELL (44)

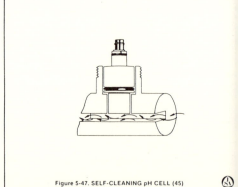

Figure 5-47. SELF-CLEANING pH CELL (45)

The reference electrode is inserted into the stream where pH is being measured as the second conductor to complete the electrical circuit. It is designed to be insensitive to all ions. It can be compared to the cold (reference) junction of the thermocouple circuit. The most reliable and reproducible reference electrode is one in which a silver-silver chloride couple contacts a saturated potassium chloride solution, which, in turn, flows through a small opening into the process stream. The silver-silver chloride electrode consists of a silver wire coated with silver chloride and immersed in the potassium chloride solution. Electrical contact between the potassium chloride solution and the process is through the porous top of the electrode (32, 33).

The measuring circuit must be capable of measuring very small voltage potentials (millivolts) and must have a very high input resistance to measure accurately the output of the electrodes. The high input resistance is necessary to limit current flow, since the current flow multiplied by the electrode resistance results in a voltage divider effect and represents an error in measuring the generated voltage (32).

The necessity of a flow through the porous plug in the reference electrode to complete the electrical circuit has proved troublesome. Flow stoppages due to the coating of the porous plug or insufficient pressure within the electrode to maintain flow require frequent maintenance. Pluggage is minimized by the solid-state electrode design, which requires no electrolyte flow. The potassium chloride and silver chloride crystals are sealed in a cell, and process liquid is permitted to enter the cell through a wooden or ceramic porous plug. The liquid enters the reference chamber and dissolves some of the salt to form a conductive path between the silver-silver chloride half cell and the liquid being measured. Although the porous plug can still become fouled, it has substantially more surface area than the small opening used to permit electrolyte more flow into the process and thus is not as readily blocked completely (27, 32, 34).

The development of the solid-state reference electrode led to sustained efforts to make the pH sensor more rugged. The original electrodes were fragile and protruded into the conduit or vessel. Therefore they were subject to breakage and fouling. Alternative designs include recessed sensing elements (44), flush sensing elements that are scrubbed clean (45), and pH-sensitive enamel on steel pipe (46). Electronic preamplifiers are mounted very close to the sensing electrodes in order to avoid transmitting the very small millivoltage generated by the measuring electrode. Encapsulating the electronics in epoxy is one technique for protecting them from the process liquid (44). Devices for cleaning the electrodes include ultrasonic cleaners, which apply high-frequency energy to the electrodes, and physical cleaners that range from nylon brushes to carborundum stones (47).

      b. Turbulent flow scrubs surface

      c. Ranges — 10° to 100°C; 100 psig maximum

   3. High-pressure, high temperature cells (see Fig. 5-48)

      a. Enamel measuring electrode on pipe surface

      b. Glass disc reference electrode

      c. Ranges — 0° to 140°C — 15 to 150 psia

C. Applications of pH measurement

   1. Process control — product quality

   2. Safe disposal of wastes

   3. Water treatment

   4. Biomedical studies

   5. Food industry

   6. Chemical processing

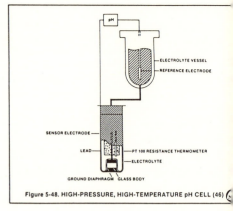

Figure 5-48. HIGH-PRESSURE, HIGH-TEMPERATURE pH CELL (46)

VII. Oxidation-reduction (redox) potential measurement

  A. Definition of oxidation and reduction — extent to which a chemical (ion) exchanges electrical charges during reaction

    1. Measure of potential difference developed due to quantity of oxidant and reductant in solution

    2. A measure of concentration ratios

    3. Determinant for control of additional reagents to a process

  B. Method of redox measurement (see Fig. 5-50)

    1. Redox cell (see Fig. 5-51)

      a. Construction

        (1) Measuring electrode — metal (platinum or gold)

        (2) Reference electrode — calomel, silver chloride

    2. Units of measurement

      a. Eh

      b. rH

  C. Factors affecting redox measurement (potential)

    1. pH of the solution (degree of ionization not sufficiently stable)

    2. Temperature

      a. Change in degree of ionization with increasing temperatures

      b. Electrode voltage a function of temperature

  D. Applications of redox measurement

    1. Extraction of chrome from industrial wastes

    2. Production of hypochlorites

    3. Indigo dye process — textile industries

    4. Copper etching (ferric chloride) — printed circuitry

    5. Alkaline chlorination process — cyanide destruction in plant wastes

    6. Organic reactions

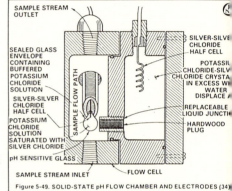

Figure 5-49. SOLID-STATE pH FLOW CHAMBER AND ELECTRODES (34)

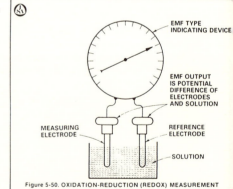

Figure 5-50. OXIDATION-REDUCTION (REDOX) MEASUREMENT

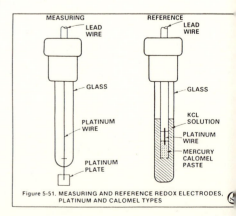

Figure 5-51. MEASURING AND REFERENCE REDOX ELECTRODES, PLATINUM AND CALOMEL TYPES

Oxygen-Reduction Potential Measurement: The oxygen-reduction potential (commonly termed ORP or Redox) is also a non-specific or inferential measurement and is distinctly similar to the pH measurement.

Many chemical reactions take place by the transfer of electrons from one substance to another. In each case, one substance is reduced by gaining one or more electrons, while the other is oxidized by losing the same electrons. Thus, the available electrons from the oxidized substance are taken up by the reduced substance until some equilibrium condition is reached (32, 35).

The relative tendency of different substances to lose electrons (the relative oxidation potentials) varies with the number of electrons on the outer shell and the size of the atom or ion. Accordingly, these substances can be arranged in a descending order of relative potentials, where the arbitrary standard for the potentials is the hydrogen electrode. The state of the reaction is then measured by the potential developed between an inert, noble metal electrode and a reference electrode.

The measuring electrode, usually gold or platinum, donates and accepts electrons. It will thus acquire the electrochemical potential of the electrons, relative to the prevailing redox equilibrium of the process solution.

# REFERENCES

1. Snowden, F.C., "Instrumentation in Pollution Control," Newsletter of the South Jersey ISA Section, September, 1971, p. 6.
2. Carroll, G.C., *Industrial Process Measuring Instruments*, McGraw-Hill Book Co., 1962.
3. Simms, R.J., "Industrial Turbidity Measurement," *ISA Transactions*, Vol. 11, No. 2, p. 146.
4. Hach, C.C., "Basic Turbidity Instrumentation," *Instruments and Control Systems,* December, 1968, p. 73.
5. Bailey Meter Co., "Turbidity Transmitter," Product specification E74-1, 1966.
6. Hallikainen, K.E., "Viscosity Measurement," *Instruments and Control Systems,* February, 1962, p. 137.
7. Hallikainen, K.E., "Viscometry," *Instruments and Control Systems,* November, 1967, p. 82.
8. Merrill, E.W., "Basic Problems in the Viscometry of Non-Newtonian Fluids," *ISA Journal,* October, 1955, p. 462.
9. Anon., "Viscometer Survey," *Instruments and Control Systems,* January, 1971, p. 66.
10. Eubank, P.T., and B.F. Fort, "The use of Rotational Viscometers with Non-Newtonian Liquids," *ISA Transactions,* Vol. 6, No. 4, p. 298.
11. Bagdon, K.M., "Viscometry in Oil Burner Control," *Instrumentation Technology,* February, 1972, p. 43.
12. Wotring, A.W., and T.B. McAveeney, "Applying Ultrasonic Viscometers to Polymer Processes," *ISA Journal,* October, 1960. p. 67.
13. Fischer & Porter Catalog/72, p. B77.
14. Tye, R.P., "The Art of Measuring Thermal Conductivity," *Instrumentation Technology,* March, 1969, p. 45.
15. Coyle, H.F., "Innovations in Furnace Atmosphere Controls," *Instruments and Control Systems,* October 1968, p. 79.
16. Lehrer, E., "Dynamic Behavior of Instruments for Gas Analysis in Connection with Automatic Control," Paper No. 75-59, 14th Annual ISA Conference, Chicago, Illinois, 1959.
17. Lavigne, J.R., "Consistency Control Concepts — Old and New," *The Journal of the Technical Association of the Pulp and Paper Industry,* Vol. 51, No. 1, January, 1968.
18. Kidder, R.J., and R. Rosenthal, "Recent Advances in Conductivity Measurement," *Instruments and Control Systems,* April, 1965, p. 118.
19. Foxboro Co., "Electrolytic Conductivity System," Bulletin K124A, 1970.
20. Bollinger, L.E., "Transducers for Measurement," *ISA Journal,* December, 1964, p. 55.
21. Foxboro Co., "Conductivity Cells," Technical Information Sheet No. 43-10a.
22. Clements, W.C., and K.B. Schnelle, "Electrical Conductivity in Dynamic Testing," *ISA Journal,* July, 1963, p. 63.
23. Young, I.G., "Conductivity: Danger, Handle with Care," ISA Paper No. 73-721, International Conference and Exhibit, October, 1973.
24. Negus, R.W., "Chemical Analysis by Inferential Techniques," *Instruments and Control Systems,* August, 1964, p. 87.
25. Ranson, J.B., "Real-Time Oxygen Measurement for Combustion Control," *Controls and Instrumentation*(British), September, 1972, p. 46.
26. Snowden, F.C., "Instrumentation in Pollution Control," Newsletter of the ISA South Jersey Section, September, 1971.
27. Smith, D.E., and F.H. Zimmerli, *Electrochemical Methods of Process Analysis* Instrument Society of America, 1972.
28. Weiss, M.D., "Electrochemical Analysis in Process Control," *ISA Journal,* January, 1961, p. 62.
29. Westinghouse Electric Corp., "Hagan Probe Type Oxygen Analyzing System," Bulletin 106-101, 1972.
30. Beckman Instruments, Inc., "An Introduction to pH," Bulletin 7222.
31. Aronson, M., "pH," *Instruments and Control Systems,* August, 1965, p. 81.
32. Shinskey, F.G., *pH and pION Control* John Wiley & Sons, New York, 1973, p. 2.
33. Moore, F.E., "Taking Errors out of pH Measurement by Grounding and Shielding," *ISA Journal,* February, 1966, p. 60.
34. Barben, T.R., "Industrial pH Measurement," *Instruments and Control Systems,* August, 1968, p. 85.
35. Jones, R.H., "Oxidation-Reduction Potential Measurement," *ISA Journal,* November, 1966, p. 40.
36. Hach, C.C., "Introduction to Turbidity Measurement," Hach Chemical Co., Ames, Iowa, 1974.
37. Condrashoff, G., "Turbidity and Suspended Solids Measurements," *Measurements and Control,* February, 1981, p. 129.
38. Condrashoff, G. "Waste Water In-Line Turbidity and Suspended Solids Measurements," Proceedings of 1979 ISA Symposium, Wilmington, Delaware, p. 239.
39. Clack, P.J. and F.L. Williams, "Turbidity and Suspended Solids: What's the Difference?" *Pollution Engineering,* March, 1981, p. 43.
40. Wright, J.V., "Inductive Method of Electrodeless Conductivity Measurement," Proceedings of 1979 ISA Symposium, Wilmington, Delaware, p. 233.
41. Foxboro Co., "Series 1210 Balsbaugh Electrodeless Conductivity Systems," PSS6-3B1A, 1979.
42. Hall, J., "More on Measurements," *Instruments and Control Systems,* July, 1981, p. 45.
43. Moron, Z., "Differential, Three Electrode Measurement of Electrolytic Conductivity," *J. Physical E.; Scientific Instrumentation* (British) 1981, p. 686.
44. GLI Bulletin S60P/177, "Model 60 Probe Assembly," Great Lakes Instruments, Inc., 7552 N. Teutonia Avenue, Milwaukee, WI 53209
45. Sensurex Bulletin 406, "Self-cleaning pH Flow System," Sensurex, 9713 Bolsa Avenue, Westminister, California 92683.
46. Bulletin 273-2e, "Pfaudler Sensor Type 03," Sybron/Pfaudler, 96 Ames St., Rochester, N.Y.
47. Bulletin 35.02e, "ORP/pH Probe Type 8322 with Mechanical Cleaning," Polymetron, 1990 S. Sproul Road, Broomall, PA 19008.

# 6
# RECORDERS

## INTRODUCTION

Recorders automatically and continuously draw a graph of measured variables versus time. The resulting graph provides a permanent record that is used to review causes for off-specification product; detect long term changes in the performance of the unit operations; assess dynamic interrelationships between unit operations; predict immediate future happenings from experience in viewing former recordings; and provide a signature from which control system performance can be optimized.

Early recorders were large, bulky devices mounted near the process, and into which process fluids were piped directly. Typically, pressure, temperature, and flow were recorded. The round chart revolved once every 24 hours and had to be changed daily. Later models provided for 7-day and 30-day chart rotations. The round chart has the additional disadvantage that chart time graduations are radial lines starting at the center so that time graduations are very close together at the center and widely spaced at the periphery.

The full-case recorder was moved from field to central control room as plant operation from a central point was introduced. Pneumatic signals in tubes were received from the measuring transmitters and transduced into a motion by bellows and Bourdon tubes. Control room investment costs almost immediately forced a reduction in size to the miniature recorder with its strip chart, which runs for a month without changing.

The miniature recorder maintained its physical characteristics but was actuated electronically to accommodate the newer electronic control systems. Electrical actuation demanded increased complexity because of such phenomena as interference pick-up by the transmitting wires; common mode voltages; circuit loading and impedance matching; and amplifier drift. However, excellent component designs are available that largely overcome these difficulties.

The process recorder has reached its zenith and is beginning to decline. It is costly to maintain: Charts must be purchased, changed, and stored; pen inking and maintenance requires attention; and panel board space in a well-lighted and air conditioned room is a premium cost. Modern recording is done within the memory of the control computer and is called up to the display on a cathode ray tube (CRT) upon demand. An attached copier then provides a paper graph of only those recordings of interest (40).

The high performance recorder is seldom permanently installed but is connected temporarily to selected process points where recordings of higher dynamic fidelity are required. The recorder charts can be run at selected speeds and are capable of very high speeds. The pens are extremely quick and can move at the speeds required to follow the fastest variable changes. In fact, the "pen" becomes a light beam in many such recorders, since the inertia of the pen is an impediment in tracking.

Digital Indicators and Recorders: Digital instruments convert analog inputs into numerical data, and indicate, log, print, or record the data upon command. Digital indicators are digital voltmeters (DVMs) and are gaining wide acceptance because of the ease of accurately reading several significant figures. Analog indicators still find their place, however, since the operator familiar with the scale can glance at it and know whether it is "right" or not by the position of the pointer. Thus actual readings are not necessary in an emergency.

The digital data is in the form of *binary arithmetic*. The binary arithmetic, which is used for counting, is a numbering system using a base number or "radix" of two. In this system there are only two digits, 1 and 0, which represent "on" and "off" conditions. The binary system is a base-2 numbering system, which operates much like the familiar base-10 numbering system. For example, to evaluate 5286:

$$(5) (10^3) + (2) (10^2) + (8) (10) + 6$$

Likewise to evaluate 1010 in the binary system:

$$(1) (2^3) + (0) (2^2) + (1) (2) + (0) (1) =$$
$$(1) (8) + (1) (2) = 10$$

A table of decimal and binary numbers follows:

| Binary | Decimal |
|--------|---------|
| 0 | 0 |
| 1 | 1 |
| 10 | 2 |
| 11 | 3 |
| 100 | 4 |
| 101 | 5 |
| 110 | 6 |
| 111 | 7 |
| 1000 | 8 |
| 1001 | 9 |
| 1010 | 10 |

Note the pattern wherein the 1 is shifted to the left and a 0 is added and replaced by a 1 for each successive number. For higher numbers, the binary number is increased by powers of 2, as shown in the table. For example, the decimal number 100 is 1100100 in binary.

To simplify the binary system, a modified form of digits called binary coded decimal (BCD) is used in numerical counters. In BCD a set of four binary numbers is used for each decimal digit. For example:

| BCD | Decimal |
|-----|---------|
| 0000 | 0 |
| 0001 | 1 |
| 0010 | 2 |
| 0011 | 3 |
| 0100 | 4 |
| 0101 | 5 |
| 0110 | 6 |
| 0111 | 7 |
| 1000 | 8 |
| 1001 | 9 |

For higher numbers using two, three, or four decimals, each BCD set is coded separately and presented in sequence. Thus, decimal 28 is 0010/1000 and 999 is 1001/1001/1001 in BCD.

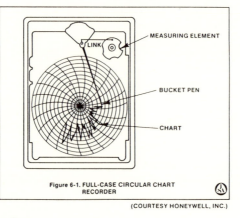

Figure 6-1. FULL-CASE CIRCULAR CHART RECORDER

(COURTESY HONEYWELL, INC.)

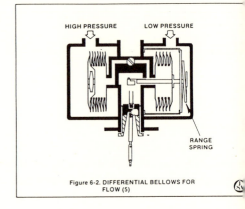

Figure 6-2. DIFFERENTIAL BELLOWS FOR FLOW (5)

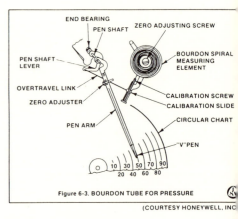

Figure 6-3. BOURDON TUBE FOR PRESSURE

(COURTESY HONEYWELL, INC

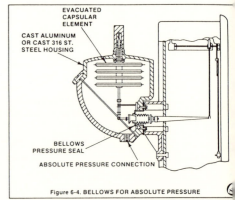

Figure 6-4. BELLOWS FOR ABSOLUTE PRESSURE

(COURTESY BRISTOL BABCO

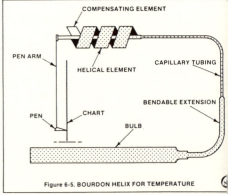

Figure 6-5. BOURDON HELIX FOR TEMPERATURE

(COURTESY BRISTOL BABCO

I. Process recorders
   A. Full-case recorders
      1. Construction (see Fig. 6-1)
         a. Weatherproof
         b. Chart drive
            (1) Spring-wound clock
            (2) Electrical
         c. Sensing element
            (1) Ring balance for flow
               (a) Pressure differential displaces mercury
               (b) Unbalance of weight of mercury causes ring to turn
               (c) Pen movement proportional to ring turning angle
            (2) Differential bellows for flow (see Fig. 6-2)
               (a) Pressure differential causes force unbalance
               (b) Contained liquid moves from one bellows to other
               (c) Force is balanced by calibrated range spring
               (d) Movement is transmitted to pen linkage
            (3) Bourdon tube for pressure (see Fig. 6-3)
               (a) Internal pressure increases
               (b) Bourdon uncoils with increasing pressure
               (c) Movement of end of Bourdon tube actuates pen
            (4) Bellows for absolute pressure (see Fig. 6-4)
               (a) Evacuated bellows in pressure-tight housing
               (b) Measured pressure acts on bellows, causing movement
               (c) Resulting reaction is transmitted to recording pens
            (5) Bourdon helix for temperature (see Fig. 6-5)
               (a) Temperature change causes mercury

The process-actuated recorder is a *full-case instrument* that must be mounted near the process where the measurements are being made since process fluids must be piped directly into the recorder. The case is of aluminum or glass-fiber-reinforced polyester construction with a sturdy, gasketed door for operation with minimum protection against the elements. Chart actuation can be by a spring-wound chart motor, which provides complete independence from external sources of power, or by an electrical motor (42).

*Sensing Elements:* The variable of interest is measured by bringing the process fluid into an element that transduces the variable into a motion. Differential pressure measuring flow is sensed by a ring-balance device, as discussed under "Secondary Element-Differential Pressure Meter" in the section on flow and by a differential bellows assembly. Pressure is sensed by a Bourdon tube (1), vacuum by an evacuated bellows (2), and temperature by a Bourdon tube with a captive fluid (3, 6). In each case, the movement of the sensing element actuates a linkage that imparts angular motion to a pen arm. Pen arm movement is calibrated to variable changes by adjusting the linkage (4).

Inking pens are used on process recorders. Up to four pens, and thus four graphs, can be mounted on a single recorder.

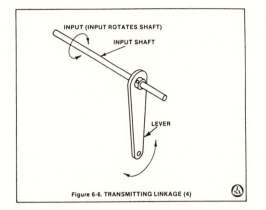

Figure 6-6. TRANSMITTING LINKAGE (4)

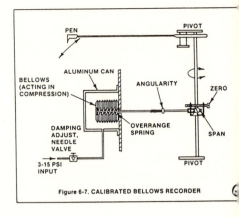

Figure 6-7. CALIBRATED BELLOWS RECORDER

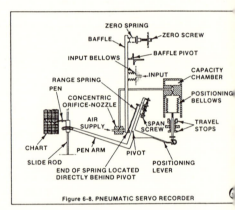

Figure 6-8. PNEUMATIC SERVO RECORDER

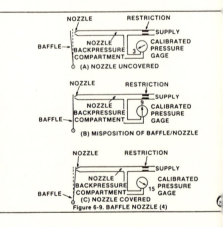

Figure 6-9. BAFFLE NOZZLE (4)

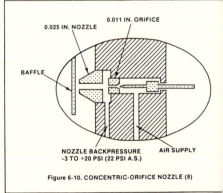

Figure 6-10. CONCENTRIC-ORIFICE NOZZLE (8)

(COURTESY SYBRON/TAYLOR)

        volume change
- (b) Volume change causes helix to deflect
- (c) Deflection is transmitted to recording pen
  d. Transmission linkage from sensor to pen (see Fig. 6-6)
    (1) Principle of operation
      (a) Sensor causes shaft to rotate
      (b) Link rotates with shaft
      (c) Link motion is transmitted to pen

II. Pneumatic recorders
  A. Calibrated bellows (see Fig. 6-7)
    1. Principle of operation
      a. Input signal damping
      b. Input signal to outside of receiver bellows balanced by internal spring
      c. Bellows motion transmitted by linkage
      d. Linkage connected to pen
  B. Pneumatic servomechanism (see Fig. 6-8)
    1. Principle of operation
      a. Input signal to input bellows
      b. Force beam changes nozzle-baffle relationship
      c. Backpressure on nozzle changes pressure in positioning bellows
      d. Positioning bellows causes movement of positioning lever
      e. Positioning lever movement changes tension on range spring
      f. Range spring tension and bellows pressure rebalance mechanism
      g. Positioning lever moves pen
    2. Baffle-nozzle (see Figs. 6-9 and 6-10)
      a. Baffle movement for 3-15 psi
        (1) 0.002 inch (Fig. 6-9)
        (2) 0.001 inch (Fig. 6-10)

The *pneumatic recorder* is simplified somewhat because the sensing element always receives a 3 to 15 psig air signal independent of the variable being recorded. The sensor, located at the process, has an integral pneumatic pilot relay capable of transmitting a pneumatic signal proportional to the variable being measured over distances of hundreds of feet through 1/4-inch tubing. Thus the controller, since it is in a protected environment, does not have to be as sturdily constructed. Early pneumatic recorders were duplicates of the process recorders with the sensor bellows and indoor construction.

The simplest pneumatic recorder has a *calibrated bellows* as does the process recorder. The pneumatic input signal first goes through a needle valve restrictor, which provides adjustable damping to filter out noise in the transmitted air signal. The damped signal then enters an airtight aluminum can where it acts on a bellows working in compression. Linearity is improved by the use of a large area bellows with the compression force counteracted by a spring. Linear spring movement is transmitted to the rotating shaft by an arbor linkage arrangement. The linkage includes adjustments for zero, span, and linearity. The recording pen is attached to the rotating shaft.

The calibrated bellows has been largely replaced by a *pneumatic servomechanism* in many miniature pneumatic recorders. The pneumatic analog signal representing the process variable actuates an input bellows (7, 8, 9). A change in input pressure acting on the bellows causes a change in the balance of forces on a force beam, resulting in a beam movement. The beam acts as the baffle or flapper in a flapper-nozzle device. A very small movement of the force beam (baffle) results in a major change in the backpressure on the nozzle. The backpressure on the nozzle is transmitted to a bellows or diaphragm that repositions an output lever. The movement of the output lever changes the tension on a range spring, which simultaneously changes the force balances on the force beam and the output lever. Both lever and beam continue to move until balance is attained.

Since the force beam is the baffle for the nozzle, it moves only a few thousandths of an inch, thus minimizing interaction between the span and zero adjustments. The output lever, on the other hand, moves sufficiently to cause the pen to move across the chart width.

*Miniature Recorders:* The transmission of signals made possible the operation of a plant from a central control room. Economy and convenience for the control room operator dictated a much higher instrument density on the panel board. The portion of the recorder visible from the front of the panel was reduced in size to 6 inches, or even 3 inches by 6 inches, and was designed so that units could be installed immediately adjacent to each other in rows. The volume necessary to contain the mechanism requires increasing extension behind the panel from 6 inches to about 20 inches.

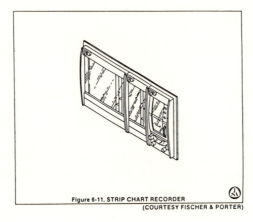

Figure 6-11. STRIP CHART RECORDER
(COURTESY FISCHER & PORTER)

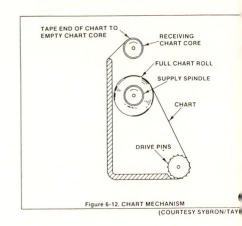

Figure 6-12. CHART MECHANISM
(COURTESY SYBRON/TAY

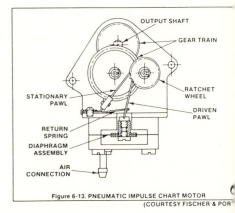

Figure 6-13. PNEUMATIC IMPULSE CHART MOTOR
(COURTESY FISCHER & POR

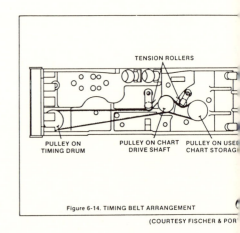

Figure 6-14. TIMING BELT ARRANGEMENT
(COURTESY FISCHER & POR

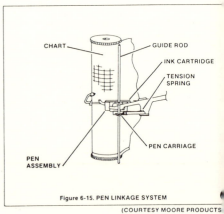

Figure 6-15. PEN LINKAGE SYSTEM
(COURTESY MOORE PRODUCTS

     (3)  Signal gain of 10,000
     (4)  Air blast on baffle introduces nonlinearity
     (5)  Small nozzle to reduce blast causes slow response

C.  Strip chart recorders (see Fig. 6-11)
  1.  3 in. × 6 in. or 6 in. × 6 in. dimensions
  2.  High density mounting
  3.  Strip chart mechanism (see Fig. 6-12)
    a.  4 inches wide
    b.  30 days duration
    c.  8 to 15 hours visible by withdrawing chassis
  4.  Chart drive
    a.  Electrical motor
    b.  Pneumatic impulse motor (see Fig. 6-13)
      (1)  4 pulses/minute
      (2)  Pulse moves diaphragm assembly upward
      (3)  Pawl moves ratchet wheel
      (4)  Pawl drops to next ratchet tooth
      (5)  Return spring returns diaphragm to null
    c.  Timing belts (see Fig. 6-14)
      (1)  Driven by shaft
      (2)  Turn timing drum
      (3)  Turn used chart shaft
    d.  Inking system
      (1)  Pens (see Fig. 6-15)
        (a)  Capillary (see Fig. 6-16)
        (b)  Fiber tip (see Fig. 6-17)
      (2)  Ink capsule (see Fig. 6-18)
        (a)  One for each color
        (b)  Replaceable
        (c)  12 month supply
        (d)  Ink will harden with age

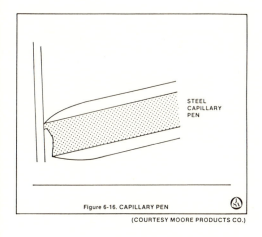

Figure 6-16. CAPILLARY PEN

(COURTESY MOORE PRODUCTS CO.)

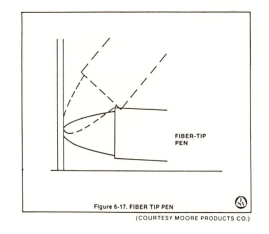

Figure 6-17. FIBER TIP PEN

(COURTESY MOORE PRODUCTS CO.)

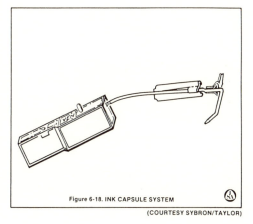

Figure 6-18. INK CAPSULE SYSTEM

(COURTESY SYBRON/TAYLOR)

Miniaturization led to the use of a long rectangular chart that is transferred from one spindle to another as the recording is made. The 4-inch wide chart is in the form of a long strip of chart, resulting in the designation *strip chart recorder*. A single chart will accommodate some 30 days of recording, and the last 7 to 9 hours will be visible by pulling the recorder from its case. The chart must be pulled from its rewind spool to review older information. A high-speed, low-torque electrical rewind motor is available, which permits the chart to be easily pulled from the spool, but which rapidly rewinds upon release. Another form of chart is the fan fold chart, which is folded accordian style. The chart moves down vertically over the drive spindle, and refolds in a receptacle below the recording pens. The folded chart is easily reviewed, refolded and replaced in the receptacle.

The *chart drive* motor is mounted on the chassis immediately behind the chart spindles. The motor is connected to the driving spindle through nylon gears and flexible timing belts. Single- or dual-speed, 110- or 220-volt, electrical drive motors are used. Alternatively a single-speed pneumatic impulse chart motor is available for hazardous environments. The pneumatic motor consists of a timing mechanism that produces 4 timed 20 psig pulses per minute. The pulses cause a ratchet gear to step a given amount, thus driving the gear proportionately to time (42). Mechanical clock drives are seldom found in strip chart recorders.

The strip chart recorder provides from 1 to 4 recording pens. Pens employing an *inking system* are widely used on pneumatic recorders. Ink is drawn through the pen to the recorder chart by the movement of the chart past the pen. The capillary-type pen, as its name implies, draws ink through the pen by capillary action. Thus it is very sensitive to the tension holding the pen against the paper and the angle of contact. The simplest pen is the capillary type with a small reservoir adjacent to the pen. Maintaining a constant flow of ink during rapid pen movement and during filling is extremely difficult. The pre-filled disposable ink cartridge, connected to the pen by flexible tubing, has proved to be a convenient and reliable system (41).

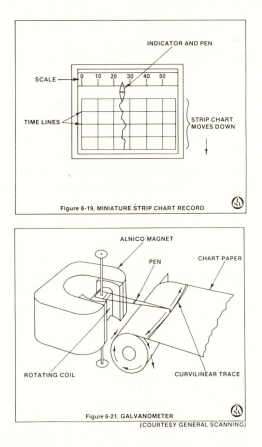

Figure 6-19. MINIATURE STRIP CHART RECORD

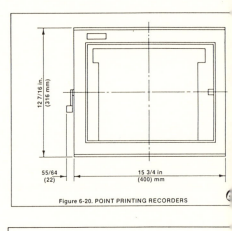

12 7/16 in. (316 mm)

55/64 (22)    15 3/4 in (400) mm

Figure 6-20. POINT PRINTING RECORDERS

Figure 6-21. GALVANOMETER
(COURTESY GENERAL SCANNING)

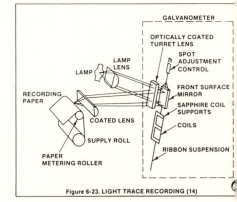

PRESSURE FLUID SUPPLY   ON—OFF VALVE

RECTILINEAR PEN LINKAGE

PENMOTOR DRIVE SIGNAL
VELOCITY/
ACCELERATION FEEDBACK

N
S

AMP

POSITION FEEDBACK   SIGNAL INPUT

Figure 6-22. GALVANOMETER WITH FEEDBACK
(14)

## III. Electronic recorders

A. Strip chart recorders — similar to pneumatic (see Fig. 6-19)

B. Point printing recorders (see Fig. 6-20)

C. Galvanometer movement (see Fig. 6-21)

1. Principle of operation
   a. Horseshoe-shaped magnet
   b. Uniform flux density across air gap
   c. Pivoted moving cell
   d. Current through coil causes rotation
   e. Coil displacement proportional to current

2. Moving iron galvanometer
   a. Fixed coil
   b. Movable magnetic element
   c. Magnet is in field established by coil
   d. Small magnetic air gap
   e. Smaller size, greater efficiency and performance

3. Velocity feedback (see Fig. 6-22)
   a. Positional feedback for linearity
   b. Velocity feedback for fast response

4. Oscillographs
   a. Portable
   b. Bandwidths of 0-60 Hz to 0-25 kHz
   c. Light trace recording (see Fig. 6-23)
   d. Thermal array recording (see Fig. 6-24)

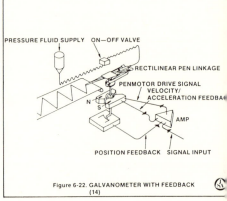

GALVANOMETER

OPTICALLY COATED
TURRET LENS

SPOT
ADJUSTMENT
CONTROL

LAMP
LENS

LAMP

RECORDING
PAPER

FRONT SURFACE
MIRROR

SAPPHIRE COIL
SUPPORTS

COATED LENS

COILS

SUPPLY ROLL

RIBBON SUSPENSION

PAPER
METERING ROLLER

Figure 6-23. LIGHT TRACE RECORDING (14)

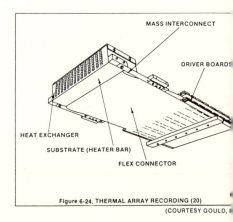

MASS INTERCONNECT

DRIVER BOARDS

HEAT EXCHANGER

SUBSTRATE (HEATER BAR)

FLEX CONNECTOR

Figure 6-24. THERMAL ARRAY RECORDING (20)
(COURTESY GOULD,

The *electronic strip chart recorder* is identical in size and appearance to the pneumatic strip chart recorder. Chart drives are similar, and many electronic recorders use the same pen and ink system. However, while distance from the field-mounted transmitter is virtually unlimited, electrical signals tend to be noisy as a result of inductive pick-up in the transmission, and recorder input filtering is sometimes required.

Electronic recorders are also similar in that they are actuated by a calibrated device, usually a galvanometer, or by a servo device, such as a potentiometric bridge circuit. The galvanometer is sensitive to the voltage drop in the transmitting wires and finds wide application in high-performance recorders that are close to the measurement. The bridge circuit is used not only in line-drawing strip chart recorders, but also in the larger, multipoint, point printing type of recorder.

*D'Arsonval Galvanometer:* The effect of electricity in motion used in the galvanometer is the force on a current-carrying conductor in a magnetic field. The D'Arsonval or permanent magnet type of actuation is a highly accurate technique for direct current measurement. The permanent magnet is usually in the form of a horseshoe and is provided with pole pieces (10, 12, 13). The pole pieces concentrate the flux from the soft iron magnet and provide a uniform flux density in the air gap. The moving coil is mounted on pivots made as frictionless as possible, which permit it to turn freely in the air gap between the pole pieces. The electrical current enters and leaves the rotating coil through two nonmagnetic spiral springs. The springs also act as a retarding force on the rotating coil. Then, when current passes through the movable coil, the coil rotates to a position such that the force exerted on the current-carrying conductors is balanced by the retarding torque of the spiral springs. Since the driving torque is directly proportional to the current and the restraining torque is directly proportional to the rotational displacement, coil displacement is directly proportional to current.

Alternatively, the *moving iron type of galvanometer* consists of a fixed coil and a movable element made of soft iron. The fixed coil carries the current, and the movable magnetic element is located in the magnetic field established by the coil. This arrangement offers small size, low weight, and ruggedness, but performance is not as good as the D'Arsonval configuration.

Feedback from the galvanometer output enhances the linearity and the speed of response of the recording pen. *Positional feedback* greatly improves linearity, and *velocity feedback* widens bandwidth to 100 cycles per second (hertz) (14, 15).

One approach to velocity feedback is to install a pick-off device on the same shaft as the torque-producing unit. Unfortunately, the arrangement requires much stiffer return springs to raise the natural frequency of the assembly above the bandwidth of the galvanometer. A more economical approach is to use the drive signal in which changes represent the velocity signal. The signal sensor is mutually coupled to the drive current, and the back emf is subtracted from the sensor output (11).

*Oscillographs* are usually portable recorders that are primarily used to record experimental test data requiring very high speeds. The galvanometer actuation is capable of bandwidths (the ability to follow a sine wave) of from 0 to 60 cycles per second (hertz) to 0 to 25,000 cycles per second (0 to 25 kHz). The feedback principles formerly discussed are utilized to attain these speeds of response (14, 15, 16).

The high speed of the stylus across the recorder chart in an oscillograph requires a very carefully designed and maintained pen and ink system and has led to the development of alternative recording mechanisms (41). One alternative is to attach a mirror to the rotating member of the galvanometer. A beam of light from a high-pressure arc lamp is collimated by a lens and is directed onto the galvanometer recording mirror. The mirror turns with the rotating member of the galvanometer, transforming the electrical current into mechanical motion. The galvanometer rotation causes the light beam to traverse the photosensitive recorder chart as it is being driven past the plane of the light beam (14, 25). The use of a mercury vapor light source provides an ultraviolet light beam (18) permitting use of direct print recorder paper, which develops dry with exposure to low level ultraviolet light.

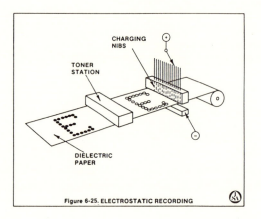

Figure 6-25. ELECTROSTATIC RECORDING

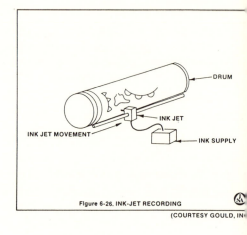

Figure 6-26. INK-JET RECORDING

(COURTESY GOULD, IN

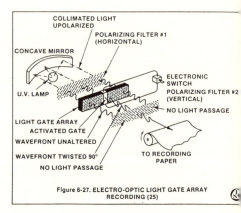

Figure 6-27. ELECTRO-OPTIC LIGHT GATE ARRAY RECORDING (25)

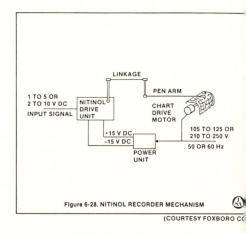

Figure 6-28. NITINOL RECORDER MECHANISM

(COURTESY FOXBORO C(

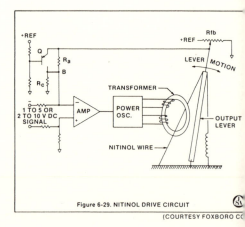

Figure 6-29. NITINOL DRIVE CIRCUIT

(COURTESY FOXBORO C(

      (1)  Styli are selectively heated
      (2)  Heat causes dot to be printed on thermal sensitive chart paper
      (3)  Paper is moved incrementally to give appearance of continuous line
  5.  Recording techniques
    a.  Heated stylus
    b.  Electric
    c.  Pressure
    d.  Electrostatic recording (see Fig. 6-25)
      (1)  High-voltage "nibs" in writing head
      (2)  Electrostatic paper charged by nibs
      (3)  Paper passes through toner
      (4)  Black and white image
    e.  Ink jet (see Fig. 6-26)
      (1)  Pressurized ink source
      (2)  Jet stream of ink on paper
      (3)  Nozzle mounted on carriage
      (4)  Nozzle and chart movements coordinated
      (5)  Quick drying ink
    f.  Thermal array recording
    g.  Electro-optical array recording (see Fig. 6-27)
      (1)  PLZT optically polarized
      (2)  Voltage application rotates plane
      (3)  Performs as light gate
      (4)  Will not pass light with no voltage
      (5)  Passes light with voltage
    h.  Electrodynamometer — actuated recorders
  D.  Nitinol® drive unit (see Fig. 6-28)
    1.  Principle of operation (see Fig. 6-29)
      a.  2 to 10 volts signal to differential amplifier
      b.  Amplifier output to oscillator
      c.  Oscillator output to transformer coil
      d.  Transformer output induces current in Nitinol wire

*Heated Stylus Recording:* A heated stylus is used with thermally sensitive recorder charts (16) up to some 90 cycles per second (hertz) bandwidth. Heat control is provided for adjustment of the width of the recorded line. The heated stylus melts or chemically affects a special temperature-sensitive coating on the paper (42).

Recording techniques also include (22):

- *Electric* — Electric discharge burns a record in a specially coated paper.

- *Pressure* — The stylus removes a wax-like coating from the paper, exposing a darker color underneath.

- *Electrostatic* — This method uses an electrical charge to write on the paper. Electrostatic paper is charged by the writing head and then passed through a toning and fixing station where an oppositely charged toner is attracted to the paper. This results in a black-and-white image. The writing head consists of many small wires or "nibs" mounted in a dielectric material and separated from each other to prevent shorting and to maintain alignment. Each nib is charged with a high voltage in the proper sequence to print an electrostatic image in the form of a tiny black dot on the paper when it emerges from the toner station. High resolution is attained by as many as 200 nibs per inch. The electrostatic method is silent and can print numerical data as well as diagrams, curves, and pictures. With the exception of some very expensive printers, however, this method is not as fast as the ink-jet system because of the necessity for the toning process.

- *Ink Jet* — A jet of quick-drying ink is sprayed on the paper through a very small nozzle as the paper moves past. The nozzle is mounted on a carriage and moved, like the moving ball on a typewriter, across the paper, which is mounted on a drum. The recorder can be designed so that either the ink jet is held stationary until the paper has made one complete revolution or the paper is held stationary until the ink jet has moved the entire length of the drum. At this point one or the other is advanced for the next sweep, and the process is repeated until the entire plot or diagram is completed. Color can be added by means of three ink jets using the three primary colors. Because the ink dries very fast as it strikes the paper, no further processing is required.

Another recording technique is *thermal array recording* (19). The thermal array has no moving parts but uses heat styli to put an analog trace on thermal-sensitive paper. The styli are arranged in a linear array and are heated by individual drivers that are selected and latched. Upon selection, the styli to be driven are heated simultaneously by an applied voltage. The process of heating selected styli can be repeated up to 200 times per second, and the paper is moved by an increment of less than the width of one element for each heating cycle. In this way the series of individual dots made by the heated element on the thermal-sensitive paper are merged into what appears to be a continuous line (49).

*Electro-optic oscillograph recorders* have been developed from later developments in electro-optic materials. One such material is lead lanthanum zirconate titanate (PLZT), which has the property of being optically isotropic when no voltage field is applied, but the application of voltages causes a virtual rotation of the plane of the polarized light. Thus the PLZT is essentially an electronic light gate with a switching time in the order of microseconds. In use, a mercury arc lamp emits unpolarized light, which is polarized linearly, proceeds through the PLTZ (which has no applied voltage) and is blocked by an analyzing polarizer. Applying voltage to the PLZT causes it to rotate the incoming plane of polarization so that it is in the same plane as the analyzing polarizer, and the light is passed through. In the application to an oscillographic recorder, a linear array of light gates transmits light through a focusing lens to a direct print paper, which moves past the recording array (14, 15, 20, 21, 48).

The *electrodynamometer* is a direct-acting electrical recorder used for the measurement of ac voltage, ac current, and power. Its principal parts are two fixed coils enclosing one movable coil. The movable coil rotates within the fixed outer coils. The same current flows through all three coils, and the current enters and leaves the movable coil through two nonmagnetic spiral springs. The rotational torque is proportional to the square of the current flow through the coils. The recorder chart must have square root graduations with scale divisions very close together at the lower end to accommodate the squared relationship. The inertia of the moving coil limits this recorder to measurements that change slowly.

The Foxboro *Nitinol*® drive unit is a unique servomechanism used to drive recorder pens (23, 24). The drive unit transforms a 0 to 10 V input signal into a pen arm rotation directly proportional to the signal.

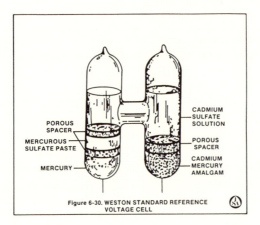

Figure 6-30. WESTON STANDARD REFERENCE VOLTAGE CELL

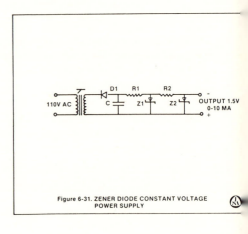

Figure 6-31. ZENER DIODE CONSTANT VOLTAGE POWER SUPPLY

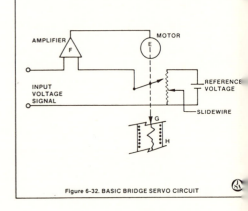

Figure 6-32. BASIC BRIDGE SERVO CIRCUIT

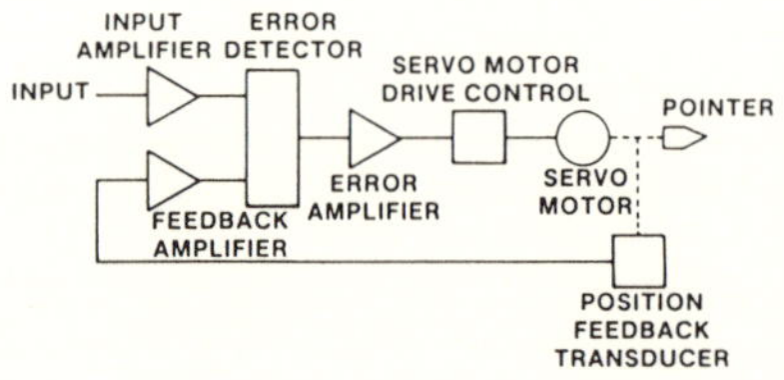

e. Current heats wire
f. Wire lengthens with increasing temperature
g. Change in length positions recorder pen and feedback slidewire wiper

E. Bridge-type servomechanisms
1. Highly sensitive
2. Null-balance
3. No voltage drop across input signal wires
4. Voltage compared to reference voltage
   a. Standard cell
   b. Zener diode compensated power supply (see Fig. 6-31)
      (1) Step down transformer T
      (2) Half-wave rectification by diode D1
      (3) Noise filtering by capacitor C
      (4) Temperature sensitive resistor R2
5. Principle of operation (see Fig. 6-32)
   a. Amplifier detects error between input voltage and feedback voltage
   b. Error drives motor, which moves wiper and recording pen
   c. Wiper moves until input voltage equals reference voltage
   d. Components (see Fig. 6-33)
   e. DC/AC converter (see Fig. 6-34)

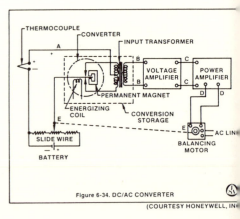

Figure 6-33. BASIC COMPONENTS
(COURTESY SYBRON/TAYLO

Figure 6-34. DC/AC CONVERTER
(COURTESY HONEYWELL, IN

The 0 to 10 volt signal to the recorder connects to a differential amplifier input. The voltage is compared to a voltage from the feedback slidewire with the wiper arm connected to the recorder pen linkage. The error output of the amplifier drives a power oscillator, which in turn feeds the primary windings of toroidal transformer. The transformer induces current into the Nitinol wire at a frequency proportional to that of the transformer.

The unusual feature of the Nitinol wire is a high sensitivity of wire length to temperature. The wire is under tensile stress, and the induced current heats the wire and causes a change in length. The change in the length of the wire causes movement of an output lever. The lever positions the wiper on the slidewire, providing the feedback signal and positioning the pen linkage. Overheating of the Nitinol wire is prevented by a transistor circuit that increases feedback voltage to the amplifier and limits amplifier output.

The *bridge-type servomechanism* is widely used to actuate null-balance recorders. It has the advantages of (1) high sensitivity, (2) independence of the resistance of the connecting wires, and (3) high accuracy. The sensitivity is due to the inherent amplification of the bridge itself and the ability to further amplify the bridge unbalance. The independence to lead-in wire resistance accrues from the null-balance operation, wherein no current flows at balance and no voltage drop appears across the wires (13, 26, 27). The bridge indicates the equality or inequality between the voltage to be measured and a very accurately known voltage.

Early null detectors used *standard voltage cells* such as that developed by Dr. Edward Weston in 1893. The standard cell contains a saturated solution of cadmium sulfate and two electrodes, one of mercury and the other a cadmium-mercury amalgam, separated by porous spacers. The cell provides 1.018636 volts at 20°C and is still used as a Primary Standard in calibration laboratories. Each cell is traceable to the National Bureau of Standards.

The standard cell, for practical reasons, was replaced by an unsaturated normal cell. However, the use of the chemicals inherent in the cell and the necessity of checking (standardizing) the operating battery against the standard cell periodically have led to the development of highly regulated and stable power supplies to replace the batteries and the standard cell (13, 28, 29, 30).

Modern power supplies use a semiconductor device called a *Zener diode*, basically a silicon crystal diode that operates to maintain a constant voltage while the current varies through it. A diode is a two-element device that will conduct when a dc voltage is applied in the forward direction but will not conduct current when the voltage is applied in the reverse direction. Reversing the direction of the positive voltage in a Zener diode causes very little current to flow until the voltage reaches a certain value called the *Zener voltage*. At this voltage the resistance of the diode breaks down and current increases rapidly. Because very large changes in current through the Zener diode cause very small changes in the voltage applied to it, it is a voltage regulator. When Zener diodes are cascaded, very high voltage regulation can be achieved. This minimizes the effects of the 60-Hz supply voltage variations so that the output from a Zener diode-regulated power supply is very stable (see Fig. 6-31).

The functional components of the bridge-type servomechanism recorder are an input conditioning module, an error detection device (the bridge itself), an error amplifier, a power amplifier to drive the servo motor, a reversing servo motor, and a calibrated variable resistor that is positioned by the servo motor and transmits a voltage back to the error detector (31, 32, 33). Because of its function, the resistor is generally called a retransmitting or repositioning slidewire. The input module is typically a plug-in printed circuit board that conditions the input with dynamic filtering circuits to reduce signal noise and provide signal conversion, amplification, and linearization (34). The error detector compares the input voltage to the voltage from the repositioning slidewire, and the error amplifier causes the servo motor to reposition to reduce the error to zero.

*DC/AC Conversion:* The output of the dc error amplifier is typically converted ("chopped") to ac by circuits in the input of the power amplifier. The converter is a flat metal reed that is driven by a magnetic coil between two contacts connected to the primary winding of the input transformer. The unbalanced error potential appears across the vibrating reed of the converter and the center tap of the input transformer. Thus, as the oscillating reed moves from one contact to the other, the error voltage causes current to flow, first in one half of the primary winding and then in the opposite direction through the other half of the primary winding. This causes an alternating current flux in the transformer core, which induces an alternating voltage in the secondary winding.

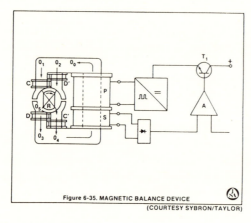

Figure 6-35. MAGNETIC BALANCE DEVICE
(COURTESY SYBRON/TAYLOR)

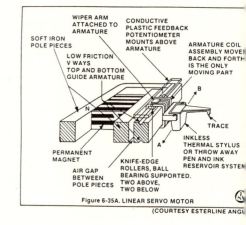

Figure 6-35A. LINEAR SERVO MOTOR
(COURTESY ESTERLINE ANGU...

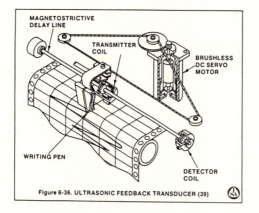

Figure 6-36. ULTRASONIC FEEDBACK TRANSDUCER (39)

f. Inductive transducer unit (see Fig. 6-35)
   (1) Replaces slidewire
   (2) Square wave is applied to coil P
   (3) Magnetic flux in core sensed by coil S
   (4) Rotor R connected to recorder pen
   (5) Rotor position sets fluxes
   (6) Fluxes induce voltages in coils
   (7) Voltage summation is error signal proportional to rotor rotation
   (8) Voltage generated in sensing coil is rectified and sent to amplifier
g. Linear servo motor (see Fig. 6-35A)
   (1) Motor field expanded linearly
   (2) Permanent magnet establishes polarity
   (3) Armature moves linearly with pen
h. Ultrasonic transducer (see Fig. 6-36)
   (1) Ultrasonic transmitter emits pulse
   (2) Propagated in magnetostrictive delay line
   (3) Sensed by detectors at ends of slidewire
   (4) Measures distances as ratios of propagation times

The oscillating reed is driven by a magnetic coil connected to the 60 Hz line through a low-voltage transformer. Thus it will oscillate 60 times per second. When the coil is energized, the tip of the reed is polarized alternately north and south by a permanent magnet. The reed is attracted to the magnet pole of opposite polarity and caused to oscillate in harmonic motion at 60 Hz frequency. The current flowing alternately in each half of the primary winding, caused by the action of the converter, therefore will induce 60 Hz alternating voltage in the secondary winding of the transformer.

*Balancing Servo Motor*: The power amplifier is an ac device to minimize errors caused by amplifier drift. The signal from the power amplifier is applied to the control winding of a reversible, two-phase balancing servo motor, which is directly coupled to the slidewire voltage pickoff arm ("wiper") through the mechanical linkage of the recording pen or the printhead assembly. The servo motor, being sensitive to the phase of the amplified error signal, drives the balancing system in the direction to reduce the error to zero. The dc motor is also being used to drive the wiper arm because it is smaller and has faster response than the ac motor. The basic circuit for driving the motor is the same except that the output of the power amplifier is converted to dc, which drives the motor in either direction depending upon the polarity of the dc output voltage.

A later development is the linear dc motor, basically an expanded field motor using a permanent magnet mounted next to the pen carriage so that it drives the pen arm directly. Although it has the advantage that no gears and cables are required, the special magnet is heavy and more expensive than a smaller geared dc motor.

Bridge-type servo recorders can use any one of the three types of motors, each of which has certain advantages suited to particular applications. The following table shows the relative advantages and disadvantages of each type.

|  | Advantages | Disadvantages |
|---|---|---|
| 1. Two-phase ac | Very reliable | Slow Speed |
|  | No brush contacts | Large size |
|  | Infinite stall | High current required |
|  | Simple amplifier |  |
| 2. DC | Fast response | Brush contacts |
|  | Low current drive | Current must be cut off or limited electrically at stall point |
|  | Low cost | Some have cogging action |
|  | Good reliability | Amplifier circuit is more complicated |
| 3. Linear dc | Very fast | Heavy magnet |
|  | No gearing | Expensive to manufacture |
|  | No brush contacts |  |

Both ac and dc motors have reduction gears that slow down the shaft speed. This increases the torque and permits balancing action to change slowly (preventing overshoot) and provides sufficient power to drive the indicating pointer and pen at a constant speed. Most self-balancing potentiometers will oscillate because of motor inertia as the null-balance point is approached unless damping is provided. Therefore, an adjustable resistor and capacitor are connected across the input to reduce instability or hunting. If the resistor is properly adjusted and the capacitor is the correct size, the balancing action can be made to approximate an exponential curve so that the motor slows down as it approaches the null-balance point, thus preventing oscillation.

The *repositioning slidewire* is a major maintenance item in the long-term operation of the bridge system. Conceptually it is a wire-wound resistor with a sliding pickoff arm that permits current to flow between the powered end of the resistor and the contact point. Thus it "picks off" the voltage represented by the resistance and current combination at that point. Difficulties include dirt on the slidewire, noise generated by the slidewire-wiper junction, heat generated at the junction, and wear on the resistor wire. These concerns have led to the development of plastic resistors, (36), an ultrasonic transducer (39), a magnetic balancing system that has no sliding contacts (35), and an inductive transducer unit (31). The inductive unit is a multi-path rotating transformer actuated by a cam mechanically connected to the servo motor output and the recording pen.

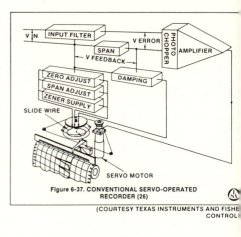

Figure 6-37. CONVENTIONAL SERVO-OPERATED RECORDER (26)

(COURTESY TEXAS INSTRUMENTS AND FISHE CONTROL

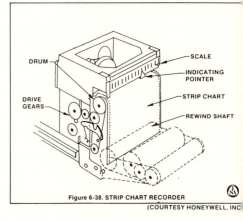

Figure 6-38. STRIP CHART RECORDER

(COURTESY HONEYWELL, INC

F. Multipoint recorders
1. Balanced-bridge mechanism
   a. Strip chart recorder configuration
      (1) Balancing motor turns pulley (see Fig. 6-37)
      (2) Cable and pulley system
      (3) Cable moves wiper, pointer, and pen simultaneously
      (4) Cable positions at servo null-balance position
      (5) Chart driven at speed to accommodate balancing (see Fig. 6-38)

Multipoint Recorders: Bridge-actuated recorders provide pen-type chart marking for up to three recorded signals, and multipoint chart printing for up to 24 recorded signals. The pen-type system uses inking pens similar to those described under pneumatic recorders. An alternative electrical writing system uses a low-voltage erosion technique to etch aluminum-coated paper, resulting in a contrasting black trace. The multipoint recorder uses the null-balance system to record the inputs. Each input is sampled, the measuring system nulls out at a point corresponding to the input signal, a printhead drops to the chart paper and prints a dot and a number identifying that input, and then the input selector switch is advanced to the next programmed point. The printing is controlled by an index and print module that has the following sequence of functions:

- Generates timing pulses at the selected printing rate

- Generates commands to index the chart drive

- Generates print commands to the solenoid that actuates the print actuation bar

- Desensitizes printing mechanism until servo amplifier reaches null position

- Synchronously moves print wheel and felt wheel so that ink pad will ink next numeral to be printed

- Programs the sequence of all functions for proper recorder operation

*Balanced-Bridge Recorder Mechanism:* Mechanically, motion of the balancing motor turns a drum that is attached to the same shaft as the slidewire. A cable wound around the drum and over a series of pulleys moves the pointer and pen to indicate and record the position of slidewire contact, which is directly related to the value of the input. In circular chart recorders the pointer is at the center of the chart, and the scale is around the outside edge. The pointer moves clockwise, and the pen carriage moves from the center to the outside edge (37). Since the diameter of the chart is usually 10 to 12 inches (25 to 30 cm), the recording pen can only move over a maximum radius of 5 or 6 inches (12.5 to 15.0 cm). The actual travel is even less because of the center diameter required to allow for the pointer shaft and hub. Charts and scales are usually specified according to the minimum scale division. For example, the minimum scale division for a range of 0 to 100 deg C is 5 deg C. For a much wider range of 0 to 500 deg C, the minimum scale division is 25 deg C. It is because of this limitation in the resolution that most bridge-type servo recorders are now of a strip chart configuration.

A strip-chart recorder with a chart width of 10 inches (25 cm) allows the pen to travel over the full width of the chart. Therefore, for a range of 0 to 500 deg C, the minimum division can be as small as 5 deg C. Thus, the resolution of the record is five times better when strip charts are used instead of circular charts. Since the measurement accuracy of most potentiometer recorders is plus or minus 1/4 percent, the actual accuracy of the record is 2.5 deg C.

The strip chart recorder has a roll of paper wound on a spool mounted on a spindle directly under the indicating pointer. The pointer and pen carriage move on a guide bar driven by a cable that moves from left to right. The motion is linear over the full width of the range. The chart, which moves down from top to bottom over a flat plate, or platen, is driven by a separate chart motor, which turns a sprocket at the top. The chart has punched holes on each side, which engage the sprocket so that it cannot slip. A separate shaft at the bottom rewinds the chart on a spool at the bottom. Some charts are made so they fold over (fan fold) at the bottom instead of rewinding, thus eliminating the rewind spool (38).

The *speed of the chart* is controlled by the rpm of the chart motor and by the gear ratios that drive the sprocket. For example, for pen speeds of 5 seconds full scale a chart speed of 2 inches/hour (5.1 cm) is usually used. However, for higher speeds, when the pen may have a response time of one second full scale, the chart speed must be increased four times to prevent the record from being smeared. It is necessary, therefore, when choosing a strip chart recorder, to consider the speed of response required to match the rate of change of the input signal and then to choose a chart speed fast enough to prevent smearing. A good rule to follow is to increase the chart speed by a factor of two for each doubling of the response time. The table below is a guide to follow for this purpose.

| Pen Response Speed | Chart Speed |
|---|---|
| 10 seconds | 2.5 cm/hr (1.0 in.) |
| 5 seconds | 5.1 cm/hr (2.0 in.) |
| 2-1/2 seconds | 10.2 cm/hr (4.0 in.) |
| 1 second | 20.4 cm/hr (8.0 in.) |

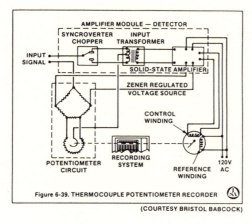

Figure 6-39. THERMOCOUPLE POTENTIOMETER RECORDER

(COURTESY BRISTOL BABCOCK)

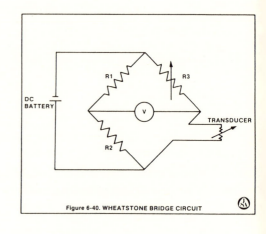

Figure 6-40. WHEATSTONE BRIDGE CIRCUIT

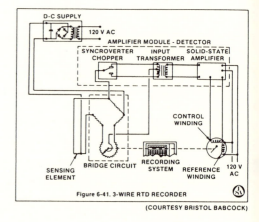

Figure 6-41. 3-WIRE RTD RECORDER

(COURTESY BRISTOL BABCOCK)

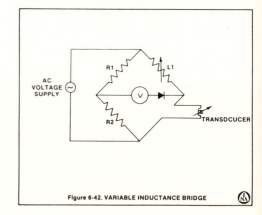

Figure 6-42. VARIABLE INDUCTANCE BRIDGE

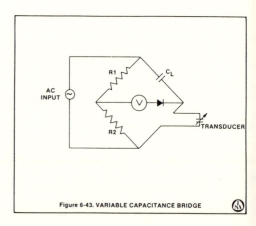

Figure 6-43. VARIABLE CAPACITANCE BRIDGE

2. Thermocouple potentiometer recorder (see Fig. 6-39)
3. DC bridge recorder
   a. Wheatstone bridge circuit (see Fig. 6-40)
   b. 3-wire RTD recorder (see Fig. 6-41)
   c. Detector resistance changes with temperature
   d. Resistance is measured as voltage drop across the detector
   e. Constant current supplied by power supply
   f. Change in voltage is detected and balanced by bridge circuit
   g. Voltage drop across lead wires cancels out
4. Resistance, temperature detectors
   a. Variable inductance bridge (see Fig. 6-42)
      (1) Unknown inductance from transducer
      (2) Balanced by variable inductance
      (3) AC voltage supply
   b. Variable capacitance bridge (see Fig. 6-43)
      (1) Unknown capacitance from transducer
      (2) Balanced by variable capacitance
      (3) AC voltage supply

Strip charts are printed with vertical and horizontal lines. The vertical lines correspond to the range of the recorder, and the horizontal lines are evenly spaced to correspond to the chart speed in inches or centimeters per unit time. To prevent errors in printing and calibration, the chart must be printed on high-quality paper. Also, because the paper will expand and contract as the ambient humidity and temperature change, recalibration of the recorder may be required. For this reason the perforations on the left edge are usually slotted so that the chart can expand or contract.

The changing of chart speed is accomplished by a single-speed electrical motor with a gearbox for two- or three-speed operation. Up to 10 different speeds can be attained by a 10-speed stepping motor, with the stepping rate set by the frequency of pulses from an oscillator circuit (29).

*Potentiometers:* The voltage input to the bridge-type recorder can come from many measuring devices such as thermocouples, pH amplifiers, photocells, and tachometers. A recorder that is compatible with voltage inputs is commonly called a potentiometer, even though this terminology conflicts with that of the variable resistor, which is also called a potentiometer.

*Thermocouple Reference Junction Compensation:* The thermocouple potentiometer recorder, however, must accommodate the differential rather than the absolute nature of the measurement. As discussed under "Thermocouples" in the temperature measurement section, the reference or cold junction of the thermocouple must be maintained at a constant, known temperature in order to calibrate the measured temperature in absolute units. In the laboratory, the reference junction is placed in an ice bath, providing a very convenient reference temperature of 0°C. The ice bath is physically impossible in an operating recorder and is replaced by a temperature-compensated reference junction. The connection of thermocouple lead-in wire to copper wire is the reference junction. This junction is embedded in a resistor wound with temperature-sensitive wire. The resistance is a direct function of temperature, and the resistor becomes part of the bridge circuit. Thus it automatically adjusts the balance of the circuit to compensate for changes in reference junction temperature (33, 36).

*Wheatstone (dc) Bridge:* The balanced-bridge recorder can be readily converted from the voltage-measuring potentiometer to the measurement of any electrical signal. The recorder is designated as follows for the alternative measurements:

• Wheatstone or dc bridge — The null-balance dc bridge is designed to measure resistance with a high degree of accuracy and sensitivity (33). It finds application with any sensing element where the resistance changes as a function of the measured variable change. Typical sensing elements are resistance temperature detectors (RTDs), strain-gage pressure transducers, and electrolytic conductivity measurements.

*Resistance temperature detectors* were described in the chapter entitled "Temperature Measurement." The measurement of temperature using a resistance detector depends upon a change in resistance, which is measured as a voltage difference across the detector caused by a small current flowing through the detector. In a three-wire resistance detector a constant current is supplied from a battery or voltage supply in the measuring circuit. The external resistance detector is connected in the bridge circuit in place of one of the resistors. A change in voltage, caused by a change in temperature, is detected and balanced by the balancing motor, which moves the slidewire contact to the null point. The detector current returns through the third wire to the battery or power supply. In this way, because the potentials across the lead wires cancel out, the resistance of the lead wires does not affect the measured voltage. No current flows in the detector lead wire because it is opposed by the slidewire potential. Because the bridge resistor is much higher in value than the detector resistance, small changes in current through the detector do not greatly affect voltage.

In general, resistance temperature detectors (RTDs) are used instead of thermocouples when narrow spans and higher accuracy measurements are required. Although the detectors cannot be used without damage for temperatures higher than 500 deg C, some manufacturers, upon special order, can supply them for higher temperatures.

*Variable Inductance Bridge* — A magnetic core attached to a sensing element such as a bellows or a diaphragm and moving linearly within a coil varies the inductance of the coil. One such device is a linear variable differential transformer (LVDT). The change in inductance is a measure of the change in the sensing device and is measured by a null-balance inductance bridge.

*Variable Capacitance Bridge* — The movement of diaphragms that cause the movement of

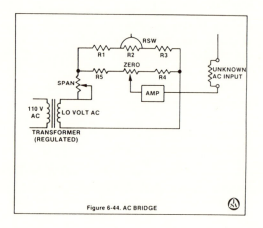

Figure 6-44. AC BRIDGE

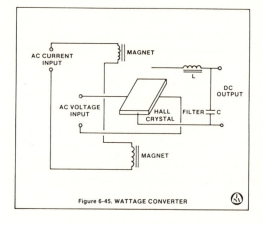

Figure 6-45. WATTAGE CONVERTER

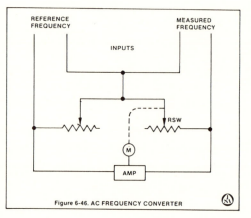

Figure 6-46. AC FREQUENCY CONVERTER

c. AC bridge (see Fig. 6-44)
  (1) Regulated transformer power supply
  (2) Measured voltage applied to slidewire and to amplifier
  (3) Measured voltage balanced against regulated voltage
d. Wattage converter (see Fig. 6-45)
  (1) Uses "Hall effect" semiconductor crystal
  (2) Input ac voltage drives current through crystal
  (3) Input ac current induces magnetic field
  (4) Current flow at right angles to magnetic field
  (5) Input current is multiplied by input voltage
  (6) Charge differential on crystal surface is proportional to wattage
  (7) Charge differential produces dc current output for potentiometer
e. AC frequency converter (see Fig. 6-46)
  (1) Comparator circuit
  (2) Reference frequency from crystal-controlled oscillator
  (3) Compared to measured frequency
  (4) Duration in frequency detected and recorded
G. Electronic recorders (see Fig. 6-47)
  1. Feedback servomechanism
  2. Similar in appearance to pneumatic recorders
  3. Input signal conditioning module (see Fig. 6-48)
    a. Scale measurement signal
    b. Amplification
    c. Limit span
    d. Dynamic (RC) filter

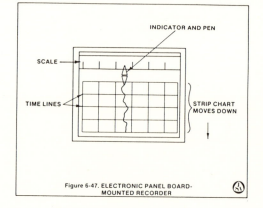

Figure 6-47. ELECTRONIC PANEL BOARD-MOUNTED RECORDER

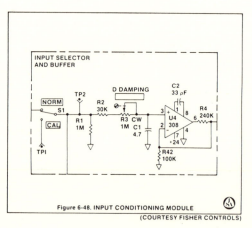

Figure 6-48. INPUT CONDITIONING MODULE
(COURTESY FISHER CONTROLS)

the plates of a condenser will change the capacitance between the plates when they are separated by a dielectric material. One of the vortex sensors described as a part of the vortex-shedding flowmeter is an example of a variable capacitance measurement, and the variable capacitance can be transduced into a milliamperage or a millivoltage by a capacitance bridge.

*AC Bridge* — The balanced-bridge servo can be used to measure ac voltage, current, power, or frequency. AC voltage or current is measured by changing the potentiometer measuring circuit to an ac bridge. The bridge voltage is obtained from a low-voltage transformer, which is applied directly across the potentiometer measuring circuit. The unknown ac input voltage is applied to the slidewire and to a special amplifier. Then it is compared to the reference voltage from the transformer. To measure ac watts or reactive power (VARs), a special converter is usually used. Of the several types, two of the more popular are the converter and the "Hall effect" transducer.

The thermal converter uses heater coils through which currents from potential and current transformers in the ac line are passed. Because ac watts are proportional to the product of voltage and current, the heat produced in the coils is directly related to the wattage. The heat produced in the coils is measured by thermocouples connected in series to form a thermopile.

The "Hall effect" converter uses a semiconductor crystal mounted between the poles of a permanent magnet. Application of a current at right angles to the magnetic field generates a charge differential on the surface of the crystal in a direction that is mutually perpendicular to both the magnetic field and the current through the crystal. DC output is obtained and is proportional to ac watts.

AC frequency can be measured by frequency converters, digital counters, and comparator circuits. An ac comparator circuit is used in a potentiometer to measure and record 60-Hz line frequency. A reference frequency from a stable crystal-controlled oscillator is applied and compared to the 60-Hz line frequency. Any deviation of 0.1 Hz or less is detected and recorded on the strip chart balanced-bridge recorder.

The *electronic panel board-mounted recorder* utilizes many of the concepts of the balanced-bridge servo recorder but is a traditional feedback servomechanism in configuration. The chart handling, chart drive, and chart marking aspects are quite similar to those discussed as a part of pneumatic panel board mounted recorders, and in fact, the electronic recorder is very similar in appearance to the pneumatic recorder when viewed from the front of the panel.

*Input Signal Conditioning*: The recorder typically accepts an input signal of 1 to 5 dc volts. An input conditioning module is thus required to:

- Scale the measurement signal to the required input by a voltage divider
- Amplify the scaled signal
- Provide high and low limits on the input
- Dynamically filter the input with an adjustable RC network

The input module can be supplied with a switch which connects the input of the buffer amplifier to a terminal, thus expediting the calibration of the recorder.

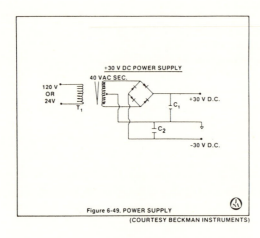

Figure 6-49. POWER SUPPLY
(COURTESY BECKMAN INSTRUMENTS)

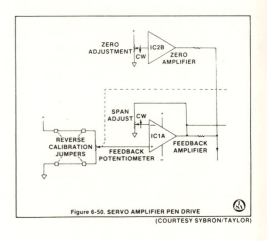

Figure 6-50. SERVO AMPLIFIER PEN DRIVE
(COURTESY SYBRON/TAYLOR)

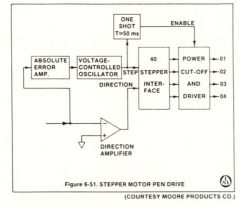

Figure 6-51. STEPPER MOTOR PEN DRIVE
(COURTESY MOORE PRODUCTS CO.)

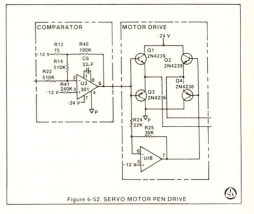

Figure 6-52. SERVO MOTOR PEN DRIVE

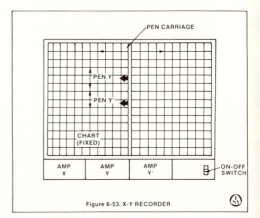

Figure 6-53. X-Y RECORDER

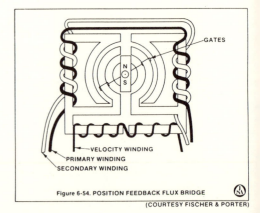

Figure 6-54. POSITION FEEDBACK FLUX BRIDGE
(COURTESY FISCHER & PORTER)

4. Power supply (see Fig. 6-49)
   a. Required voltages ±4 to ±30 dc volts
      (1) Scale measurement
      (2) Amplification
      (3) Input span
      (4) Dynamic filtering
5. Servo amplifier (see Fig. 6-50)
   a. Inverting
   b. Zero and span adjustment
   c. Compare input and feedback
6. Pen drive
   a. Oscillator (see Fig. 6-50)
   b. Stepper motor (see Fig. 6-51)
   c. Servo motor (see Fig. 6-52)
7. Pen position feedback
   a. Slidewire with wiper
   b. Fischer and Porter flux bridge (see Fig. 6-54)
      (1) 10 kHz primary coil excitation
      (2) Secondary coil voltage proportional to gate position
      (3) Gate directly couple to torque motor shaft
      (4) Feedback proportional to shaft position
      (5) Permanent magnet acts in conjunction with velocity coil to provide velocity feedback
8. X-Y recorder (see Fig. 6-53)

*Power Supply:* Voltages required by the internal circuitry of the recorder vary from $\pm 4$ dc volts to $\pm 30$ dc volts. The 120 ac volt, 60-Hz power supply consists of a transformer, a bridge rectifier, and one or more stages of filtering and voltage regulation. The transformer steps supply voltage down to the ac voltage required by the recorder. Transformer output is converted to dc by a diode-bridge rectifier and appropriate filtering, The bridge output is the source of power for voltage regulators producing the bi-polar sources, i.e., V+ and V−, required by the solid-state circuitry. The power supply is a source for the operational amplifiers, alarm amplifiers, and voltage divider network (43, 44, 45).

*Servo Amplifier:* Stages of amplification follow the input buffering circuits. One amplifier provides signal inverting as required and zero and span adjustment for one channel of the recorder. A second amplifier is the summing of servo amplifier, which may, in practice, encompass two stages of amplification. The servo amplifier compares the input signal to the position of the recording pen and drives the servo motor to reduce the difference (error) to zero.

Stepper motors are widely used as servo motors. For this arrangement, the output of the servo amplifier serves as the input to both an oscillator amplifier and a direction amplifier. The oscillator is clock or crystal regulated and provides pulses that cause rotation of the stepper motor. An absolute error amplifier between the servo amplifier and the oscillator provides a technique for making motor speed proportional to error and for removing power from the servo motor when the error is zero (44, 46).

*Pen Drive Motor:* The direction amplifier senses the polarity of the error from the servo amplifier and causes the servo motor to drive the pen in a direction to cause the error to become zero. The direction circuit is required whether the pen-positioning motor is a stepper motor or a continuous servo motor. The pen and scale indicators are driven by the motor through a cable and drum arrangement. The drum is mounted on the motor shaft so that pen, indicator, and feedback wiper assume positions proportional to the motor shaft position (44, 47).

The *feedback slidewire* with voltage pickoff wiper is similar to that used on bridge-type servomechanisms and is subject to the same weaknesses. The weaknesses have been somewhat alleviated by the use of thick film-resistive technology, as previously discussed, but the electromechanical contact between wiper and slidewire still limits reliability. The Fischer and Porter Co. (45) has replaced the slidewire completely with a flux bridge, which is a contactless, solid-state position and velocity feedback device. The primary coils are energized from a 10 kHz oscillator. The voltage developed in the secondary coils is directly proportional to the position of a moving stator, which changes the flux distribution in the core. The moving stator is directly connected to the shaft of a torque motor with limited span of rotation which drives the pen. Hence, pen position can be calibrated directly to the recorder input signal. A permanent magnet attached to the stator operates in conjunction with a velocity coil to provide a velocity feedback signal, which is applied to the servo amplifier input to regulate pen speed.

The chart drive mechanism is similar to that used on pneumatic recorders. Stepper motors are sometimes used to drive the chart spindle, with pulses supplied by an oscillator and variable speed provided by a frequency divider circuitry.

X-Y Recorders: Conventional recorders plot one or more variables against time. The X-Y recorder can plot any variable against any other variable including time. The chart is held in place electrostatically or by a vacuum system. One variable moves a pen servo system (mounted on a carriage) vertically, while the other variable drives the carriage horizontally. Both pens are driven by circuitry similar to that described under "Bridge-Type Servomechanisms" (50). Continuously variable selectable ranges for both variables are a feature of the X-Y recorder.

The recorder requires dc inputs. The input is conditioned by dynamic filtering and amplification. The conditioned input signal is transmitted to the servo amplifier, which compares it to a voltage from the feedback slidewire. A directional amplifier senses "plus" and "minus" around one input "zero" point and drives the servo motor to reduce the error to zero (18, 51, 52).

A second pen can be added so that two variables can be plotted against the third variable. This configuration is called an X-Y-Y′ recorder. Because of the flat horizontal bed of the recorder, felt tipped pens are widely used as the writing medium.

a. Plots any one or two variables against any other variable
b. Chart held in place by vacuum or electrostatically
c. Fiber tipped pen
d. Inputs (see Fig. 6-53)
   (1) DC voltage input
   (2) Input conditioning
   (3) Servo amplifier determines difference between input and feedback
   (4) Servo motor drives pen to reduce error

9. Digital indicators and recorders
  a. Analog-to-digital conversion
    (1) Successive approximation (see Fig. 6-55)
      (a) Unknown voltage sequentially compared
      (b) Voltages for comparison generated internally
      (c) Each incremental step in generated voltage one half previous step
      (d) Values of incremental steps are summed
      (e) Approximation complete when sum equals voltage to within tolerance
      (f) Very rapid
    (2) Dual-slope integration (see Fig. 6-56)
      (a) Capacitor charged by input voltage for fixed period of time
      (b) Resulting integrated voltage returned to zero by reference voltage
      (c) Time to discharge proportional to input voltage
      (d) Accurate
      (e) Slow
  b. Mechanical printer (see Fig. 6-57)
    (1) Print wheel rotates at constant speed
    (2) Print wheel prints a number on command
    (3) Action similar to typewriter
    (4) Printed through linked ribbon
  c. Electronic thermal array printer (see Fig. 6-58)
    (1) Linear thermal elements
    (2) Each thermal element represents an amplitude to be recorded
    (3) Microprocessor selects element to be heated
    (4) Microprocessor steps paper ahead
    (5) Heated element makes dot on paper
    (6) One dot for each step becomes recording

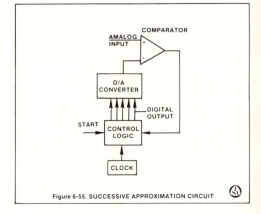

Figure 6-55. SUCCESSIVE APPROXIMATION CIRCUIT

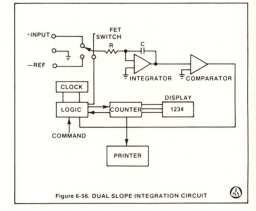

Figure 6-56. DUAL SLOPE INTEGRATION CIRCUIT

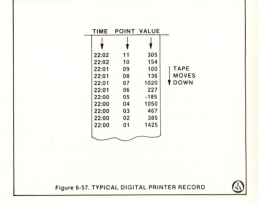

Figure 6-57. TYPICAL DIGITAL PRINTER RECORD

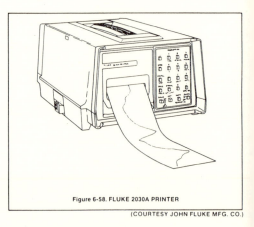

Figure 6-58. FLUKE 2030A PRINTER

(COURTESY JOHN FLUKE MFG. CO.)

*Analog-to-Digital (A/D) Conversion:* There are several ways to convert dc analog signals to BCD pulses:

- *Successive approximation* — The unknown input voltage is sequentially compared with voltage steps generated in the instrument. Until the two voltages are equal, each step is one half the value of the previous step. The values of the steps are accumulated and displayed as a decimal number equal to the input value (53).

- *Ramp and counter* — A linear rise in voltage, generated in the instrument, is compared to the unknown input value. A counter is started from a clock pulse at the start of the ramp and is stopped when the unknown and known voltages are equal at the input to a comparator. The number of pulses counted during this time is proportional to the unknown input value.

- *Dual-slope integration* — An unknown input voltage is used to charge a capacitor for a fixed period of time determined by a counter. The resulting integrated voltage is then returned to zero by a reference voltage of opposite polarity. The time required to discharge the capacitor is proportional to the average value of the unknown input voltage. During the discharge period the output of a clock-pulse generator is gated to a counter. At the end of the time period the counter operates a register to display the voltage measured, and the cycle is repeated (53).

- *Voltage-to-Frequency Conversion* — An unknown input voltage is used to generate a series of pulses whose rate is proportional to the input value. The pulses are gated to a counter for a fixed time interval. The number of pulses counted is proportional to the input value. The pulses actuate a register that displays the value as a numerical output. Should the input be from thermocouples, linearization is usually incorporated in the input amplifier by comparators, each set to a voltage corresponding to the thermocouple output curve. The gain of the input amplifier is changed so that the input to the DVM is a linear signal that is used to display the temperature directly.

*Mechanical Printer:* BCD output from the DVM or a digital thermometer can be used to actuate a printer at periodic intervals or on command whenever the measurement is to be recorded. There are a number of different types of printers available, both mechanical and electrical, which can be used for recording. The mechanical types use a paper tape and a print wheel, which is rotated at a constant speed. Upon receipt of a print command, the print wheel prints a number that corresponds to the BCD code on the tape. The action is similar to a typewriter in that the number is impressed on the paper tape through an inked ribbon at a high rate of speed. When multiple inputs are being scanned, a time signal and a point number can also be printed simultaneously to identify the record.

*Thermal Array Printer:* The electronic printer incorporates a microprocessor that provides considerable intelligence. It permits both span and limits of the digital recording to be specified. The paper tape (chart) is then divided into a 100-dot-wide field. Under the paper is a row of thermal elements in a heat sink. Each element corresponds to one amplitude of the signal to be recorded. The elements are selectively heated upon signal from the microprocessor, creating a dot on the heat-sensitive chart paper. By heating a selected junction for each step made by the paper, a graph or recording can be made. Element selection and paper advance are both under control of the microprocessor. Also, by selecting several elements each time the paper is advanced, characters are created in a row-by-row fashion (54, 55).

d. Transient recorder (see Fig. 6-59)
   (1) A/D conversion
   (2) Digital memory to store very rapid transients
   (3) External trigger for short duration transient
   (4) Analog recording from memory

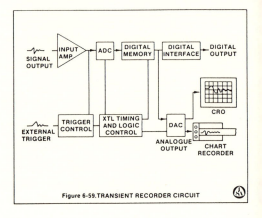

Figure 6-59. TRANSIENT RECORDER CIRCUIT

*Electronic Transient Recorder:* An advance from the electronic printer is the electronic transient recorder. This device is applied to the recording of transients lasting from 0.05 microseconds to 2 milliseconds (56). Circuitry includes the analog-to-digital converter, as formerly described, and a digital memory in which the transient is stored. The transient recorder will fill the memory to capacity upon an external trigger for planned experiments, or it can record continuously with the oldest data being discarded for random events. Sampling is then stopped at a predetermined number of samples after the external trigger. A digital-to-analog converter then provides an output for the recording.

The ink jet system is widely used in digital recorders and printers.

# REFERENCES

1. Bulletin, "Instruction on the Proper use of your Dickson Chart Recorder," The Dickson Co., 930 South Wedgewood Drive, Addison, Ill., 60101.

2. Specification Sheet G600-5a, "Series 500 Absolute Pressure Recorders," Bristol Division of ACCO, Waterbury, Conn., 06720.

3. Specification Sheet T800-5C, "Class 5 Filled System Thermometers," Briston Division of ACCO, Waterbury, Conn., 06720.

4. Patrick, D. R., *Instrumentation Training Course-Volume 1-Pneumatic Instruments*, Howard W. Sams & Co., Inc., Indianapolis, Ind., 1979.

5. Bulletin, "Process Control Instrumentation," Publication 105A 15M 4/71, The Foxboro Co., Foxboro, Mass. 02035.

6. Bulletin, "40P Series Temperature Recorders," Product Specification PSS3-1A2B, The Foxboro Co., Foxboro, Mass. 02035.

7. Instruction Bulletin, "Series 51-1330 Pneumatic Concept 45 Recorder," Bulletin 51-1330, Fischer and Porter Co., Warminster, Pa. 18974.

8. Product Data Bulletin, "Instruments, Theory, and Operation," Bulletin PDS-2M001, Taylor Instrument Process Control Division, Sybron Corp., Rochester, N.Y. 14601.

9. Service Bulletin, "Series 20000 6x6 Recorder Chassis," Service Instruction SD2000, Moore Products Co., Spring House, Pa., 19477.

10. Hubble, R., "Galvanometer Direct Writing Recorders," *Measurement and Control*, April, 1980, p. 178.

11. Montagu, J.I., and H.J. Dumas, "Miniaturization of Galvanometers and Strip Chart Recorders," *Measurements and Control*, December, 1979, p. 126.

12. Reed, H.H., and G.F. Corcoran, *Electrical Engineering Experiments*, John Wiley and Sons, 1948.

13. Patrick, D.R., *Instrumentation Training Course-Volume 2-Electronic Instruments*, H.W. Sams & Co., Inc., Indianapolis, Ind., 1979.

14. Howes, P.A., "Data Display Recording Devices," Proceedings of 25th International Instrumentation Symposium, Anaheim, Ca., 1979.

15. Bulletin HR2000/679, "CRC HR2000 Datagraph Recorder and Data Processor," CEC Division, Bell & Howell, Inc., Pasadena, Ca.

16. Bulletin 3130-379-10, "Gibson Biophysical and Industrial Recorders," Gibson Medical Electronics, Middleton, WI.

17. Bulletin 4042A, "Rustak Miniature and Servo Strip Chart Recorders," Gulton Measurement and Control Systems Division, East Greenwich, R.I.

18. Bulletin, Soltec Corp., Sun Valley, CA.

19. Miller, L., "Thermal Array Recording," *Measurements and Control*, April, 1980, p. 131.

20. Anon., "Linear Array Recorders," *Measurements and Control*, June, 1980, p. 172.

21. Samek, N.E., "A PLZT Electrooptic Shutter Array," Proceedings of 25th International Instrumentation Symposium, Anaheim, Calif., 1979.

22. Hubbard, James R., "Disposable Pens, A Revolution in Ink-Writing Systems," *Measurements and Controls*, October, 1978, p. 119.

23. Technical Information Bulletin 220-160, "SPEC 200 Nitinol Drive Unit," The Foxboro Co., Foxboro, Mass., 1973.

24. Technical Information Bulletin 39-242a, "E20S Series Electronic Consotrol Recorders," The Foxboro Co., Foxboro, Mass., 1975.

25. Anon., "Galvanometer Light Beam Recording," *Measurements and Control*, September, 1980, p. 217.

26. Anon., "Potentiometric (Servo) Recorders," *Measurements and Control*, December, 1979, p. 188.

27. Anon., "Bridges and Potentiometers," *Measurements and Control*, June, 1980, p. 151.

28. Product Data P1212-2, Bristol Division, ACCO, Waterbury, Connecticut 06720.

29. Operators Manual for Model 500 Recorder, Linear Instruments Corporation, Irvine, Calif. 92713.

30. Service and Technical Information Manual for Speedomax 165 and 250 Industrial Recorders, Leeds and Northrup Co., North Wales, Pa. 19454.

31. Instructions, Multi-Scan Recorder, Bulletin 1B-13E106, Sybron/Taylor Instrument Division, Rochester, N.Y., 1979.

32. Instruction Manual, 2800 Recorder, Molytek, Inc., Pittsburgh, Pa., 15222.

33. Bulletin ME71, Section IV, Theory of Operation, Tracor Westronics.

34. Bulletin, Chessell 10 inch Servo Recorders, Chessel Corporation, Newtown, Pa. 18940.

35. Bulletin 3142J, Potentiometric Recorder, International Products and Technologies, Inc., Jenkintown, Pa., 19046.

36. Bulletin 4E1A2-e, ER180 Recorders, Yokogawa Corporation of America, Elmsford, N.Y, 10523.

37. Bulletin, Micromax Recorders-Model R3000 Series, Leeds and Northrup Co., North Wales, Pa. 19454.

38. Bulletin, "Basic Principles of Brown Electronic Continuous Balance System," Honeywell Inc., 1100 Virginia Drive, Ft. Washington, Pa. 19034, 1956 and 1958.

39. Stegenga, J.V., "Ultrasonic Feedback Recording Potentiometer," *Measurements and Control*, June, 1980, p. 131.

40. Morris, H.M., "Trends in Recording of Process Variables," *Control Engineering*, September, 1979, p. 74.

41. Morris, H.M., "Strip Chart Recorders Keep Pace with Technology," *Control Engineering*, April, 1980, p. 63.

42. Merritt, R., "Graphic Recorder Survey," *Instruments and Control Systems*, December, 1980, p. 29.

43. Anon., "Recorders," Beckman Instruments, 2500 Harbor Blvd., Fullerton, Calif. 92634.

44. Service Instructions, Model Series 360 Electronic Recorder, Moore Products Co., Spring House, Pa. 19477, 1977.
45. Instruction Bulletin for Series 1321 Electronic Scan-Line® Recorder, Publication No. 21140, Fischer and Porter Co., Warminster, Pa. 18974.
46. Instructions for Taylor Quick-Scan® 1300 Recorders, Publication No. 1B-13A502, Sybron-Taylor, Rochester, N.Y.
47. Instruction Manual, Type RO222 Recorder, Fisher Controls Co., Marshalltown, Iowa, 1979,
48. Design Notes, "First All Solid-State Direct Writing Recorder," *Control and Instrumentation* (British), May, 1980, p. 17.
49. Feedback, "Thermal Array Recorders Fit Both Process and Lab Applications," *Control Engineering*, July, 1980, p. 8.
50. Anon:, "X-Y Recorders," *Measurements and Control*, October, 1980, p. 209.
51. Bulletin, "Handbook of Temperature Measurements," Linseis, Princeton Junction, N.J. 08550.
52. Bulletin, "1-6 Channel Flat Bed Recorder," Netzsch, Inc., Lionville, Pa., 19353.
53. Hull, M.L., "Principles of Digital Data Acquisition," *ISA Transactions*, Volume 19, No. 3, 1979, p. 25.
54. Instruction Manual, Model 2030A Programmable Printers, John Fluke Mfg. Co., Mountainlake Terrace, WA 98043.
55. Field Sales Manual, Models 2020A and 2030A PTI Printers, John Fluke Mfg. Co., Mountainlake Terrace, WA 98043.
56. Widenka, R., "Recorders with Digital Storage Capture Transients for Analysis," *Control and Instrumentation* (British), April, 1979, p. 51.
57. Johnson, T.E., "Looking for an X-Y Recorder?" *Instruments and Control Systems*, January, 1981, p. 5.

# APPENDIX
## SUGGESTED FURTHER READING

Benedict, Robert P., 1981. *Fundamentals of Temperature, Pressure, and Flow Measurements,* Third Edition. John Wiley and Sons, New York, N.Y.

Gregory, B.A., 1981. *An Introduction to Electrical Instrumentation and Measurement Systems.* John Wiley and Sons, New York, N.Y.

Johnson, Curtis D., 1981. *Process Control Instrumentation Technology.* John Wiley and Sons, New York, N.Y.

Anderson, Norman A., 1981. *Instrumentation for Process Measurement and Control,* Second Edition. Chilton Book Co., Radnor, Pa.

Rhodes, T.J. and G.C. Carroll, 1972. *Instruments for Measurement and Control.* McGraw-Hill Book Co., New York, N.Y.

Doebelin, E.O., 1975. *Measurement Systems - Application and Design,* Second Edition. McGraw-Hill Book Co., New York, N.Y.

Stein, P.K., 1962. *Measurement Engineering.* Stein Engineering Services, Tempe, Ariz.

Woodruff, J., 1964. *Basic Instrumentation.* The University of Texas, Austin, Tex.

Sweeney, R.J., 1953. *Measurement Techniques in Mechanical Engineering.* John Wiley and Sons, New York, N.Y.

Greve, J.W. (editor), 1967. *Handbook of Instrument Metrology.* Prentice Hall, Inc., Englewood Cliffs, N.J.

Carroll, G.C., 1962. *Industrial Process Measuring Instruments.* McGraw-Hill Book Co., New York, N.Y.

Holzbock, W.G., 1962. *Instruments for Measurement and Control.* Reinhold Publishing Corp., New York, N.Y.

Kirk, F.W. and N.R. Rimboi, 1975. *Instrumentation.* American Technical Society, Chicago, Ill.

O'Higgins, P.J., 1966. *Basic Instrumentation - Industrial Measurement.* McGraw-Hill Book Co., New York, N.Y.

Aronson, M.H., 1960. *Handbook of Electrical Measurements.* Instruments Publishing Co., Pittsburgh, Pa.

Thomas, H.E., 1967. *Handbook of Electrical Instruments and Measurement Techniques.* Prentice Hall, Inc., Englewood Cliffs, N.J.

Koa, Charles K., 1982. *Optical Fiber Systems.* McGraw-Hill Book Co., New York, N.Y.

Scott, R.W.W., 1982. *Developments in Flow Measurement - I.* Applied Science Publishers, London, England.

Jones, Barry E., 1977. *Instrumentation, Measurement, and Feedback.* McGraw-Hill Book Co., New York, N.Y.

Wendlandt, Wesley W., 1974. *Handbook of Commercial Scientific Instruments.* Marcel Dekker, Inc., New York, N.Y.

Westcott, C. Clark, 1978. *pH Measurements.* Academic Press, New York, N.Y.

Wilson, J.A., 1978. *Industrial Electronics and Control.* Science Research Associates, Palo Alto, Calif.

Zbar, P.B., 1972. *Industrial Electronics and Text Lab Manual,* Second Edition. McGraw-Hill Book Co., New York, N.Y.

Rhodes, T.J., 1972. *Industrial Instruments for Measurement Cont.,* Second Edition. McGraw-Hill Book Co., New York, N.Y.

Schweitzer, P.A., 1972. *Handbook of Values.* Industrial Press, New York, N.Y.

Soisson, H.E., 1975. *Instrumentation in Industry.* John Wiley and Sons, New York, N.Y.

*Standards and Practices for Instrumentation, 1980,* Sixth Edition. Instrument Society of America, Research Triangle Park, N.C.

Stearns, R.F., 1951. *Flow Measurement With Orifice Meters.* D. VanNostrand Co., Inc., Princeton, N.J.

Timmerhaus, K.D., W.J. O'Sullivan, and E.F. Hammel (editors), 1974. *Proceedings of the 13th International Conference on Low Temperature Physics - Lt 13, Vol. 4: Electronic Properties, Instrumentation and Measurement.* Plenum Press, New York, N.Y.

Lyons, J.L., 1975. *Lyons Encyclopedia of Values.* D. Van Nostrand Co., Inc., Princeton, N.J.

Mandl, M., 1961. *Industrial Control Electronics.* Prentice Hall, Inc., Englewood Cliffs, N.J.

Olsen, Lief O. and Carl Halpern, 1967. *Bibliography of Termperature Measurement, July 1960 to December 1965,* National Bureau of Standards Monograph 27, Suppl. 2 US Government Printing Office, Washington, D.C.

Preston-Thomas, H., T.P. Murray, R.L. Shepard (editors), 1972. *Temperature: Its Measurement and Control in Science and Industry, Vol. 4, Pt. 1: Basic Methods, Scales and Fixed Points, Radiation.* Instrument Society of America, Research Triangle Park, N.C.

Quinn, T.J., 1975. *Temperature Measurement Nineteen Seventy-Five.* American Institute of Physics, New York, N.Y.

Irwin, Lafayette K. (editor), 1977. *NBS Special Publication (US), Vo. 484, Pt. 2: Flow Measurement in Open Channels and Closed Conduits, Vol. 1.* US Government Printing Office, Washington, D.C.

Jones, E.B. *Instrument Technology.* Butterworth Publications, Inc., Woburn, Mass.

Jones, E.B., 1974. *Instrument Technology. Vol. 1,* Third Edition. Newnes-Butterworth, London, England.

Kaganov, I. *Industrial Applications of Electronics,* MacMillan Publishing Co., Inc., Riverside, N.J.

Kallen, H., 1961. *Handbook of Instrumentation and Controls.* McGraw-Hill Book Co., New York, N.Y.

*Evolution of the International Prac. Temperature of 1968.* American Society for Metals.

Graham A.R., 1974. *Introduction to Engineering Measurements.* Prentice Hall, Inc., Englewood Cliffs, N.J.

Hayward, A., 1979. *Flowmeters - A Basic Guide and Source Book for Users.* John Wiley and Sons, New York, N.Y.

Hewitt, G.F., 1979. *Measurement of Two Phase Flow Parameters.* Academic Press. New York, N.Y.

Hwang, Ned H.C. and Nils A. Normann (editors), 1977. *Cardiovascular Flow Dynamics and Measurements,* University Park Press, Baltimore, Md.

Durrani, Tariq S. and Clive A. Greated, 1977. *Laser Systems in Flow Measurement.* Plenum Press, New York, N.Y.

*Engineering Equipment Users Association Handbook, No. 34: Installation of Instrumentation and Process Control Systems, 1973.* Constable, London, England.

*Equivalent Values Reference Manual,* Rev. Edition, 1978, Gulf Publishing Co., Houston, Tex.

Ewing, G.W., 1971. *Topics in Chemical Instrumentation.* Journal of Chemical Education, Springfield, Pa.

Ewing, Galen W. (editor), 1977. *Topics in Chemical Instrumentation, Vol. 2.* American Chemical Society, Washington, D.C.

Ewing, Galen W. (editor), 1971. *Topics in Chemical Instrumentation; A Volume of Reprints from the Journal of Chemical Education.* Chemical Education Publishing Co., Easton, Pa.

Alfred, H. *Industrial Electronics Principles & Practice.* TAB Books, Inc., Ridge Summit, Pa.

Beckwith, T.G., 1982. *Mechanical Measurements,* Third Edition. Addison Wesley, Redding, Mass.

Beilby, Alvin L. (editor), 1970. *Modern Classics in Analytical Chemistry; a Collection of Articles from Analytical Chemistry and Chemical and Engineering News, Selected for Their Value as Supplementary Reading About Modern Chemical Instrumentation and Analytical Methods.* American Chemical Society, Washington, D.C.

Billing, B.F. and T.J. Quinn (editors), 1975. *Conference Series No. 26: Temperature Measurement,* Invited and Contributed Papers from European Conference on Temperature Measurement Held at the National Physical Laboratory, Teddington, 9-11 April 1975. Inst. Phys., London, England.

Brombacher, William G., 1967. *Bibliography and Index on Vacuum and Low Pressure Measurement, January, 1960, to December, 1965,* National Bureau of Standards Monograph 35, Suppl. 1. US Government Printing Office, Washington, D.C.

Cameron, J.F. (editor), 1970. *Neutron Moisture Gauges: A Guide Book on Theory and Practice,* International Atomic Energy Agency Technical Reports Series, No. 112. IAEA, Vienna, Austria.

Canfield, Eugene B., 1965. *Electronic Control Systems and Devices.* John Wiley and Sons, New York, N.J.

Chapman, Robert L. (editor), 1969. *Environmental Pollution Instrumentation.* Instrument Society of America, Research Triangle Park, N.C.

Cheremisinoff, Nicholas P., 1979. *Applied Fluid Flow Measurement: Fundamentals and Technology.* Marcell Dekker, Inc., New York, N.Y.

Clark, W.J., 1965. *Flow Measurement.* Pergamon, Oxford, England.

Cockrell, W.D., 1958. *Industrial Electronics Handbook.* McGraw-Hill Book Co., New York, N.Y.

Considine, D.M., 1972. *Encyclopedia of Instrumentation and Control.* McGraw-Hill Book Co., New York, N.Y.

Dadda, Luigi and Umberto Pellegrini, 1966. *Automation and Instrumentation.* Pergamon, New York, N.Y.

Kalvoda, R., 1975. *Operational Amplifiers in Chemial Instrumentation,* John Wiley and Sons, New York, N.Y.

Keast, D.N., 1967. *Measurements in Mechanical Dynamics.* McGraw-Hill Book Co., New York, N.Y.

Kinzie, P.A., 1973. *Thermocouple Temperature Measurement,* Wiley-Interscience, Division of John Wiley and Sons, New York, N.Y.

Kirk, F.W., 1975. *Instrumentation,* Third Edition. American Testing Society, Philadelphia, Pa.

Kloeffler, R.G., 1960. *Industrial Electronics and Control,* Second Edition. John Wiley and Sons, New York, N.Y.

Lavigne, John R., 1972. *An Introduction to Paper-Industry Instrumentation.* Miller Freeman, San Francisco, Calif.

*Temperature, Its Measurement and Control in Science and Industry, 1941.* American Institute of Physics, Reinhold, N.Y.

Draper, C.S., W. KcKay, and S. Lees, 1955. *Instrument Engineering,* Vol. 2 - Methods for Associating the Situations of Instrument Engineering. McGraw-Hill Book Co., New York, N.Y.

Draper, C.S.l, W. McKay, and S. Lees, 1955. *Instrument Engineering,* Vol 2 - Methods for Associating Mathematical Solutions with Common Forms. McGraw-Hill Book Co., New York, N.Y.

Jones, E.B., 1953 *Instrument Technology.* Butterworth Scientific Publications, London, England.

Baker, H.D. and E.A. Ryder, 1953., *Temperature Measurement in Engineering.* John Wiley and Sons, New York, N.Y.

Pearson, C.B., 1958. *Technology of Instrumentation.* Van Nostrand, Princeton, N.J.

Beckwith, T.G. and N.L. Buck, 1961. *Mechanical Measurements.* Addison-Wesley Publishing Co., Reading, Mass.

*Precision Measurement and Calibration, National Bureau of Standards Handbook 77,* Vol. I - Electricity and Electronics, Vol. II - Heat and Mechanics, Vol. III - Optics, Metrology and Radiation, 1961. US Government Printing Office, Washington, D.C.

Coxon, W. 1959. *Flow Measurement and Control.* The Macmillan Co. New York, N.Y.

Coxon, W. 1960. *Temperature Measurement and Control.* The Macmillan Co., New York, N.Y.

Harrison, T.R., 1960. *Radiation Pyrometry and Its Underlying Principles of Radiant Heat Transfer.* John Wiley and Sons, New York, N.Y.

Tyson, Jr., F.C., 1961. *Industrial Instrumentation.* Prentice-Hall, Englewood Cliffs, N.J.

Kallen, H.P., 1961. *Handbook for Instrumentation and Control.* McGraw-Hill Book Co., New York, N.Y.

Lion, K.S., 1959. *Instrumentation in Scientific Research - Electrical Input Transducers.* McGraw-Hill Book Co., New York, N.Y.

Rider, J.F. and S.P. Prensky, 1959. *How to Use Meters,* Second Edition. Rider.

Martinek, R.G. *Technical Characteristics of Clinical Laboratory Instruments.* American Medical Technologists, Park Ridge, Ill.

*Manual on Installation of Refinery Instruments and Control Systems,* Part I - Process Instrumentation and Control, Part II - Process Steam Analyzers, American Petroleum Institute, New York, N.Y.

*Manual on Use of Thermocouples in Temperature Measurement - STP 470.* American Society for Testing Materials, Philadelphia, Pa.

*Process Measurement and Control Terminology,* PUB 219, PMC 20-2-1970. Scientific Apparatus Makers Assn., New York, N.Y.

Spink, L.K. *Principles and Practice of Flow Meter Engineering,* Ninth Edition. Foxboro Co., Foxboro, Mass.

Cusick, C.F. (editor). *Flowmeter Engineering Handbook,* Fourth Edition. Honeywell Inc., Waltham, Mass.

Babcock, R.H. *Instrumentation and Control in Water Supply and Wastewater Control.* Water and Wasters Engg., New York, N.Y.

Liptak, B. (editor and author). *Instrument Engineers' Handbook,* (Volume I - Process Measurement, Volume II - Process Control). Chilton Book Co., Radnor, Pa.

Clark, W.J. *Flow Measurement,* (By square-edged orifice plate using corner tappings). Permagon Press, Inc., Elmsford, N.Y.

Mancy, K.H. *Instrumental Analysis for Water Pollution Control.* Ann Arbor Science Publishers, Ann Arbor, Mich.

Rothfus, R.R. *Working Concepts of Fluid Flow.* Bek Technical Publications, Bridgeville, Pa.

Summer, S.E. *Electrical Sensing Controls.* Chilton Book Co., Radnor, Pa.

Luppold, D.S. *Precision DC Measurements and Standards.* Addison-Wesley Publishing Co., Reading, Mass.

Prensky, S.D. *Advanced Electronic Instruments and Their Use.* (Hard cover, paperback), Hayden.

Prensky, S.D. *Electronic Instrumentation,* Second Edition. Hayden.

Oliver, B.M. and J.M. Cage. *Electronic Measurements and Instrumentation.* McGraw-Hill Book Co., New York, N.Y.

Norton, H.M. *Handbook of Transducers for Electronic Measuring Systems.* Prentice-Hall, Inc., Englewood Cliffs, N.J.

Hougen, J.O. *Measurements and Control Applications for Practicing Engineers.* Cahners.

Liptak, B. (editor). *Instrumentation in the Process Industries.* Chilton Book Co., Radnor, Pa.

Lavigne, J.R. *An Introduction to Paper Industry Instrumentation.* Miller Freeman (also available from Foxboro Co.).

Vanzetti, R. *Practical Applications of Infrared Techniques.* John Wiley and Sons, New York, N.J.

Kinzie, P.A. *Thermocouple Temperature Measurement.* Wiley-Interscience, Division of John Wiley and Sons, New York, N.Y.

Considine, Douglas M. (editor), 1964. *Handbook of Applied Instrumentation.* McGraw-Hill Book Co., New York, N.Y.

Fribance, Austin E., 1962. *Industrial Instrumentation Fundamentals. McGraw-Hill Book Co., New York, N.Y.*

*Instruments for Measurement and Control,* Second Edition, 1962. Reinhold Publishing Corp., New York, N.Y.

Considine, Douglas M., 1957. *Process Instruments and Controls Handbook.* McGraw-Hill Book Co., New York, N.Y.

Robertson, Don, 1953. *The Primary Elements to Measure the Variable.* Leeds and Northrup Co., North Wales, Pa.

*Pyrometry - Bulletin No. 1.* Minneapolis-Honeywell Regulator Company, Brown Instrument Division, Service Department Training School, Ft. Washington, Pa.

*Mercury Barometers and Manometers - NBC Monograph 8,* 1960. U.S. Department of Commerce, National Bureau of Standards, Washington, D.C.

Diehl, John C., 1955. *Orifice Meter Constants - Handbook E-2.* American Meter Company, Inc.

Hernandez, J.S. *Introduction to Transducers for Instrumentation.* Statham Instruments, Inc., Los Angeles, Calif.

# ISA
# INDIVIDUAL STANDARDS AND RECOMMENDED PRACTICES

| | |
|---|---|
| RP2.1 — | Manometer Tables |
| S5.1 — | (ANSI Y32.20, approved January 1975) Instrumentation Symbols and Identification |
| S7.4 — | Air Pressures for Pneumatic Controllers and Transmission Systems |
| RP12.1 — | Electrical Instruments in Hazardous Atmospheres |
| S12.4 — | Instrument Purging for Reduction of Hazardous Area Classification |
| RP16.1,2,3 — | Terminology, Dimensions, and Safety Practices for Indicating Variable Area Meters (Rotameters, Glass Tube, Metal Tube, Expansion Glass Tube) |
| RP16.4 — | Nomenclature and Terminology for Extension Type Variable Area Meters (Rotameters) |
| RP16.5 — | Installation, Operation, Maintenance Instructions for Glass Tube Variable Area Meters (Rotameters) |
| RP16.6 — | Methods and Equipment for Calibration of Variable Area Meters (Rotameters) |
| S20 — | Specification Forms for Process Measurement and Control Instruments, Primary Elements, and Control Valves |
| S26 — | (ANSI MC4.1, approved April 1975) Dynamic Response Testing of Process Control Instrumentation |
| RP31.1 — | (ANSI/ISA RP31.1, approved January 1977) Specification, Installation, and Calibration of Turbine Flowmeters |
| RP37.2 — | Guide for Specifications and Tests for Piezoelectric Acceleration Transducers for Aerospace Testing |

S37.3 — (ANSI MC6.2, approved October 1975) Specifications and Tests for Strain Gage Pressure Transducers

S37.6 — (ANSI MC6.5, approved July 1976) Specifications and Tests of Potentiometric Pressure Transducers for Aerospace Testing

S37.8 — (ANSI/ISA-1977) Specifications and Tests for Strain Gage Force Transducers

S37.10 — (ANSI MC6.4, approved October 1975) Specifications and Tests for Piezoelectric Pressure and Sound-Pressure Transducers

S37.12 — Specifications and Tests for Potentiometric Displacement Transducers

S51.1 — Process Instrumentation Terminology

MC96.1 — Temperature Measurement Thermocouples

# GOVERNMENT DOCUMENTS

Most public libraries have fairly complete information on the technical literature output of our government's agencies but specific subjects can be hard to find. For the field of instrumentation — particularly measurement —the National Bureau of Standards, Washington, D.C. 20234, is a prime source.

NBS Special Publication 305, *Publications of the National Bureau of Standards,* latest edition, is available from the Superintendent of Documents, U.S. Government Printing Office, Washington, D.C. 20402 at a reasonable cost.

Other NBS publications of interest include the NBS Special Publication 300 Series, *Precision Measurement and Calibration,* a compilation of papers by NBS authors on various topics such as Temperature (Vol. 2), Heat (Vol. 6), and Ionizing Radiation (Vol. 10).

Documents from NBS non-periodical series are available from the National Information Service, Springfield, Va. 22151.

# ADDRESSES OF PUBLISHERS

Academic Press
111 5th Avenue
New York, NY 10003

Addison-Wesley Publishing Co., Inc.
Reading, MA 01867

Ann Arbor Science Publishers, Inc.
P.O. Box 1425
Ann Arbor, MI 48106

ASTM, 1916 Race St.
Philadephia, PA 19103

Bek Technical Publications, Inc.
P.O. Box 478
Bridgeville, PA 15017

Butterworth Publications
10 Tower Office Park
Woburn, MA 01801

Chemical Publishing Co., Inc.
155 W. 19th Street
New York, NY 10011

Chilton Book Co.
Chilton Way
Radnor, PA 19089

Marcel Dekker, Inc.
270 Madison Avenue
New York, NY 10016

The Foxboro Company
Publications Division
Cocasset Building
Foxboro, MA 02035

Miller Freeman Publications
500 Howard Street
San Francisco, CA 94105

Gulf Publishing Company
Box 2608
Houston, TX 77001

Hayden Book Company
50 Essex Street
Rochelle Park, NJ 07662

Honeywell, Inc.
Industrial Division, MS 440
Fort Washington, PA 19034

Industrial Press
200 Madison Avenue
New York, NY 10157

Instruments and Control Systems
56th and Chestnut Streets
Philadelphia, PA 19139

The Macmillan Co
866 Third Avenue
New York, NY 10022

McGraw-Hill Book Co.
1221 Avenue of the Americas
New York, NY 10020

Permagon Press, Inc.
Maxwell House
Fairview Park
Elmsford, NY 10523

Plenum Publishing Corp
233 Spring Street
New York, NY 10013

Prentice-Hall, Inc.
Englewood Cliffs, NJ 07632

Reinhold Book Co.
430 Park Ave.
New York, NY 10022

Scientific Apparatus Makers Assoc.
Process Measurement and Control Section
370 Lexington Avenue
New York, NY 10017

TAB Books
Blue Ridge Summit, PA 17214

Superintendent of Documents
U.S. Government Printing Office
Washington, DC 20402

U.S. Gauge Division
AMETEK, Inc.
Sellersville, PA 18960

University Park Press
300 N Charles
Baltimore, MD 21201

D. Van Nostrand Co., Inc.
135 W. 50th St.
New York, NY 10020

Wiley-Interscience
A Division of John Wiley and Sons
605 Third Avenue
New York, NY 10158

Water and Wastes Engineering
Book Department
466 Lexington Avenue
New York, NY 10017